Foundations of Optimum Experimental Design

Mathematics and Its Applications (*East European Series*)

Andrej Pázman

Mathematical Institute, Slovak Academy of Sciences,
Bratislava, Czechoslovakia

Foundations of Optimum Experimental Design

D. Reidel Publishing Company
A MEMBER OF THE KLUWER ACADEMIC PUBLISHERS GROUP
Dordrecht / Boston / Lancaster / Tokyo, 1986

Library of Congress Cataloging in Publication Data

Pázman Andrej.
Foundations of Optimum Experimental Design.

(Mathematics and Its Applications. East European Series)
Translation of: Základy optimalizácie experimentu.
Bibliography: p. 228
Includes index.
1. Experimental Design. I. Title. II. Series:
Mathematics and Its Applications (D. Reidel Publishing Company).
East European Series.
QA279.P3913 1986 658.4'03'4 85-24331
ISBN 90-277-1865-2

Distributors for the U.S.A. and Canada
Kluwer Academic Publishers,
190 Old Derby Street, Hingham, MA 02043, U.S.A.

Distributors for Albania, Bulgaria, Chinese People's Republic,
Cuba, Czechoslovakia, German Democratic Republic, Hungary,
Korean People's Democratic Republic, Mongolia, Poland, Rumania,
U.S.S.R., Vietnam, and Yugoslavia
VEDA, Publishing House of the Slovak Academy of Sciences,
Bratislava, Czechoslovakia.

Distributors for all remaining countries
Kluwer Academic Publishers Group,
P.O. Box 322, 3300 AH Dordrecht, Holland.

Original title: *Základy optimalizácie experimentu.*
First edition published in 1980 by VEDA, Bratislava.
First English edition published in 1986 by VEDA, Bratislava,
in co-edition with D. Reidel Publishing Company, Dordrecht, Holland.

Printed in Czechoslovakia.

Table of Contents

Editor's Preface

Approach your problems from the right
end and begin with the answers. Then
one day, perhaps you will find the final
question.

'The Hermit Clad in Crane Feathers' in
R. van Gulik's *The Chinese Maze Murders.*

It isn't that they can't see the solution.
It is that they can't see the problem.

G. K. Chesterton. The Scandal of Father
Brown 'The Point of a Pin'.

Growing specialization and diversification have brought a host of monographs and textbooks on increasingly specialized topics. However, the "tree" of knowledge of mathematics and related fields does not grow only by putting forth new branches. It also happens, quite often in fact, that branches which were thought to be completely disparate are suddenly seen to be related.

Further, the kind and level of sophistication of mathematics applied in various sciences have changed drastically in recent years: measure theory is used (non-trivially) in regional and theoretical economics; algebraic geometry interacts with physics; the Minkowsky lemma, coding theory and the structure of water meet one another in packing and covering theory; quantum fields, crystal defects and mathematical programming profit from homotopy theory; Lie algebras are relevant to filtering; and prediction and electrical engineering can use Stein spaces. And in addition to this there are such new emerging subdisciplines as "completely integrable systems", "chaos, synergetics and large-scale order", which are almost impossible to fit into the existing classification schemes. They draw upon widely different sections of mathematics.

This programme, *Mathematics and Its Applications*, is devoted to such (new) interrelations as *exempli gratia*:

— a central concept which plays an important role in several different mathematical and/or scientific specialized areas;
— new applications of the results and ideas from one area of scientific endeavour into another;
— influences which the results, problems and concepts of one field of enquiry have and have had on the development of another.

The *Mathematics and Its Applications* programme tries to make available a careful selection of books which fit the philosophy outlined above. With such books, which are stimulating rather than definitive, intriguing rather than encyclopaedic, we hope to contribute something towards better communication among the practitioners in diversified fields.

Because of the wealth of scholarly research being undertaken in the Soviet Union, eastern Europe, and Japan, it was decided to devote special attention to the work emanating from these particular regions.

Thus it was decided to start three regional series under the umbrella of the main MIA programme.

Two relatively new areas in mathematics are "optimization" and "statistics" and both are well on their way to becoming almost separate disciplines. If one crosses the two one may well obtain optimal experimental design as one of the offspring. As such the topic of this book partakes of several disciplines and may well be called interdisciplinary, especially if one also looks at the mathematical tools involved which can range from pure combinatorics via all kinds of optimization mathematics (combinatorial, variational, classical, mathematical programming) to information theory. And so they do in this stimulating book about nonclassical mathematics.

The unreasonable effectiveness of mathematics in science...

Eugene Wigner

Well, if you knows of a better'ole, go to it.

Bruce Bairnsfather

What is now proved was once only imagined.

William Blake

As long as algebra and geometry proceeded along separate paths, their advance was slow and their applications limited.
But when these sciences joined company they drew from each other fresh vitality and thenceforward marched on at a rapid pace towards perfection.

Joseph Louis Lagrange

Bussum, August 1984

MICHIEL HAZEWINKEL

Preface to the English Edition

The book is intended to present the mathematical background necessary for understanding, exploiting and developing the theory of experimental design and methods for computing optimum designs. Experience with the first edition has shown that the book can be used successfully as a starting point for further research and as a textbook for students of statistics.

The second edition provides some supplementary results: the Schur ordering of designs, the use of alternative methods of estimation, and singular designs are discussed in more detail, a new method for computing D-optimum designs under two alternative models is presented, etc. On the other hand, the first and the last chapters have been slightly abbreviated. On the whole, however, the book does not differ greatly from the first edition in that the emphasis is on the systematic self-contained exposition of the linear theory of estimation and of the convex theory of experimental design.

Bratislava, July 1984 *The author*

Basic Symbols

$\equiv$	— the equality sign to define a new symbol
R^k	— the k-dimensional Euclidean space
R	— $\equiv R^1$
$\boldsymbol{u}, \boldsymbol{v}$	— vectors
u_i, resp. $\{\boldsymbol{u}\}_i$	— the i-th component of the vector $\boldsymbol{u}$
$\mathbf{1}$	— the vector with all components equal to 1
$\mathbf{A}, \mathbf{B}$	— matrices
$\{\mathbf{A}\}_{ij}$	— the i, j-th element of the matrix $\mathbf{A}$
$\{\mathbf{A}\}_{i.}$	— the i-th row of the matrix $\mathbf{A}$
$\{\mathbf{A}\}_{.j}$	— the j-th column of the matrix $\mathbf{A}$
tr $\mathbf{A}$	— the trace of $\mathbf{A}$
det $\mathbf{A}$	— the determinant of $\mathbf{A}$
$\boldsymbol{u}', \boldsymbol{A}'$	— the transpositions of $\boldsymbol{u}$ and of $\mathbf{A}$
$\mathbf{A}^{-1}$	— the matrix that is the inverse of $\mathbf{A}$
$\mathbf{A}^-, \mathbf{A}^+$	— the g-inverses of $\mathbf{A}$ (see II.3)
$\|\mathbf{A}\|, \|\mathbf{A}\|_\infty, \|\mathbf{A}\|_p$	— the g-inverses of $\mathbf{A}$ (see IV.1)
$\mathbf{A}_{\mathrm{I}}, \mathbf{A}_{\mathrm{II}}, \mathbf{A}_{\mathrm{III}}$	— submatrices of $\mathbf{A}$ (see IV.1)
$\mathbf{I}$	— the identity matrix
diag $(r_1, \ldots, r_k)$	— the $k \times k$ diagonal matrix with diagonal elements $r_1, \ldots, r_k$
$\mathcal{M}(\mathbf{A})$	— the linear space spanned by the columns of $\mathbf{A}$ (II.3)
$\mathcal{N}(\mathbf{A})$	— the zero space of $\mathbf{A}$ (II.3)
$\mathcal{R}^{k \times k}$	— the linear space of all $k \times k$ matrices
$\mathfrak{S}^{k \times k}$	— the linear space of all $k \times k$ symmetric matrices

X	— the set of all trials possible in an experiment (II.2)
$C(X)$	— the set of all functions that are continuous on X
Θ	— the set of all states of the observed object
ϑ	— a state of the observed object
$f_1, \ldots, f_m$	— a linear basis in Θ (II.3)
$\alpha_1, \ldots, \alpha_m$	— unknown regression parameters (II.3)
$\hat{\alpha}_1, \ldots, \hat{\alpha}_m$	— best unbiassed estimates ($\equiv$BLUE) for $\alpha_1, \ldots, \alpha_m$
ξ, η, etc.	— designs
ξ_x	— the design concentrated at one point x
X_ξ	— $\equiv \{x \in X: \xi(x) > 0\}$
Ξ (resp. $\bar{\Xi}$)	— the set of all designs in a regression model (or in a functional model)
E, E_ϑ, $\mathrm{E}(.\vert\vartheta)$, E_ξ, etc.	— the operators of taking mean values of random variables
D, D_ϑ, etc.	— the operators of taking variances
$\mathrm{D}(z)$	— the variance of the random variable z
$\mathbf{D}(\boldsymbol{z})$	— the covariance matrix of the random vector $\boldsymbol{z}$
cov (y, z)	— the covariance of the random variables y and z
$y(x)$	— the random variable (or its value) observed in the trial x
$y^\xi(B)$	— the random variable (or its value) observed on the set B (in the functional model)
$\sigma^2(x)$	— $\equiv \mathrm{D}[y(x)]$
g, g^*, etc.	— linear functionals defined on Θ
$\mathbf{g}, \mathbf{g}^*$, etc.	— the vectors defining the functionals g, g^*, etc. (II.2)
$\hat{g}, \hat{g}^*$, etc.	— the best unbiassed estimates (=BLUE) for g, g^*, etc.
$\mathrm{cov}_\xi\,(g, g^*)$	— the covariance of $\hat{g}$ and $\hat{g}^*$
$\mathrm{var}_\xi g$	— $\equiv \mathrm{cov}_\xi(g, g)$

$\mathbf{M}(\xi)$, $\mathbf{M}$, $\mathbf{M}$, etc.	— information matrices (II.2)
$\mathfrak{M}$	— the set of all information matrices in an experiment
$\mathfrak{M}_+$	— $\equiv\{\mathbf{M}\in\mathfrak{M}: \det \mathbf{M}\neq 0\}$
ϕ, ϕ_0, ϕ_p	— optimality criteria functions (IV.1)
$\nabla\phi(\mathbf{M})$, $\nabla_{\mathbf{M}}\phi(\mathbf{M})$	— the gradient of ϕ at the point $\mathbf{M}$ (IV.1)
$\partial\phi(\bar{\mathbf{M}}, \mathbf{M})$	— the directional derivative of ϕ at the point $\bar{\mathbf{M}}$ in the direction of $\mathbf{M}$
$\mathfrak{M}_\phi$	— $\equiv\{\mathbf{M}\in\mathfrak{M}: \phi(\mathbf{M})<\infty\}$
χ_B	— the indicator of a set B
co (Q)	— the convex hull of a set Q
S	— $\equiv$ co $(\{\mathbf{f}(x): x\in X\}\cup\{-\mathbf{f}(x): x\in X\})$
$\varkappa(K)$	— the Hilbert space with the reproducing kernel K (VII.1)
$L^2(\xi)$, $L^2(\Omega, \mathcal{S}, P)$	— L^2– spaces (VII.1)
$\langle .,. \rangle$	— the inner product in a Hilbert space

The other symbols are defined in the text.

Chapter I

Introductory Remarks about the Experiment and Its Design

This book is dealing with the mathematical background of the optimization of experiments. To approach the subject more intuitively, and to obtain a survey of the book, the book begins with by this introductory chapter.

In general, an experiment is a procedure prepared and performed to get a deeper knowledge about a given object. Observations (measurements) are an important part of this procedure. A simplified definition of an experiment, which is sufficient to understand this book, is that an experiment is a set of observations (measurements) performed with a common aim. Hence, sets of measurements that are necessary to prepare a technological project or to check the actual state of a building, are also experiments.

In general, an experiment is a complex set of activities beyond the possibilities of mathematics. It needs the preparation of materials, measuring instruments, and the organization of a team of specialists. However, in contemporary experiments the role of experimental data processing gains importance. The increase of the complexity of many experiments makes also an optimum design of experiments necessary, to gain maximum information at minimum expense, and this is to be treated mathematically. Three spheres of action of mathematics in experimental activities, namely the mathematical modelling of the experiment, data processing and optimum experimental design, are parts of an operational approach to experiments. It stresses the importance of acquiring a large amount of information from the experiment. This is a complement to the approach of the sciences, physics, chemistry, biology, emphasizing the content and interpretation of the information obtained from an experiment.

Any theory of experiments has to start with considering *mathematical models of experiments*. The model that is used in this book is *the regression model with uncorrelated observations* (Chapter II). It is a rather simple model often used in applications. No wonder that the development of optimization of experiments is connected primarily with this model. In Chapter II.4 some deviations from the model are considered, and in Chapter VII the model is generalized to *the functional model* which may be infinitely dimensional.

Because of the underlying physical aspect of the experiment, a model of an experiment must contain a description of *states of the object observed*, and a description of quantities that can be observed or computed from the estimated state. In a regression model *the set of possible trials* is specified first, and then the state is defined as a function assigning to every trial the mean value of the quantity observed in the trial. These quantities are considered random variables, which allows one to take into account the instability of the conditions of the measurements. The aim of a good design of an experiment is to reduce the influence of such instabilities as much as possible.

When the model of the experiment is specified, it is necessary to express *the purpose of the experiment* and the intentions of a future user of the experimental data in terms of the model. The general requirement that the aim of an experiment is to get maximum information is not sufficient for designing the experiment. Therefore, the theory presents different *optimality criteria* (Chapter IV), and the experimenter has to choose one of them, according to his purposes. However, the choice of the optimality criterion as well as the specification of the model need a careful confrontation of the theory with the real situation, and the practical experience of the experimenter is unavoidable here.

When the model and the aim of the experiment are specified, then methods for data processing and design can be prepared. All this is to be based on mathematics, and is considered in this book.

The efficacy of designing experiments can be illustrated on elementary examples.

EXAMPLE 1. *Weighing of three objects* [5]
Let A_1, A_2, A_3 be objects of unknown weights α_1, α_2, α_3. We have to specify α_1, α_2, α_3 by four weighings on a precise balance.

Probably everybody would choose the following design of weighing which can be said "standard".

1st weighing ... nothing on the balance (to estimate the systematic error of the balance),
2nd weighing ... the object A_1 on the balance,
3rd weighing ... the object A_2 on the balance,
4th weighing ... the object A_3 on the balance.

Let y_0, y_1, y_2, y_3 denote the data obtained on the balance successively. Obviously, *the estimates* of the weights are

$$\hat{\alpha}_i = y_i - y_0; \quad (i=1, 2, 3). \tag{1}$$

Let α_0 denote the unknown systematic error of the balance and ε_i the random error of the i-th weighing. Evidently

$$y_0 = \alpha_0 + \varepsilon_0,$$
$$y_i = \alpha_i + \alpha_0 + \varepsilon_i; \quad (i=1, 2, 3). \tag{2}$$

The mean values of the random errors ε_i are zero ($E(\varepsilon_i)=0$), their variances are constant ($D(\varepsilon_i)=\sigma^2$), and they are independent. Therefore,

$$E(y_0)=\alpha_0, \; E(y_i)=\alpha_i+\alpha_0; \quad (i=1, 2, 3) \tag{3}$$

and

$$D(y_i)=\sigma^2, \; \operatorname{cov}(y_i, y_j)=0; \quad (i, j=0, 1, 2, 3, \; i\neq j). \tag{4}$$

Hence, it follows from Eq. (1) that

$$E(\hat{\alpha}_i)=\alpha_i, \; D(\hat{\alpha}_i)=D(y_i)+D(y_0)=2\sigma^2; \quad (i=1, 2, 3). \tag{5}$$

Now, let us consider another design which is said to be "rational".

1st weighing ... three objects on the balance,
2nd weighing ... A_1 on the balance,
3rd weighing ... A_2 on the balance,
4th weighing ... A_3 on the balance.

Results of the weighings are denoted by y_0, y_1, y_2, y_3 again. Instead of Eqs. (3) we have now

$$\begin{aligned} E(y_0) &= \alpha_0 + \alpha_1 + \alpha_2 + \alpha_3, \\ E(y_i) &= \alpha_0 + \alpha_i; \quad (i = 1, 2, 3). \end{aligned} \tag{6}$$

Instead of the estimates (1) we have

$$\hat{\alpha}_i = \frac{y_0 + y_i - y_j - y_k}{2} \tag{7}$$

$$(i = 1, 2, 3, j \neq k, \ i \neq j, i \neq k).$$

Hence, instead of Eqs. (5)

$$\begin{aligned} &E(\hat{\alpha}_i) = \alpha_i, \\ &D(\hat{\alpha}_i) = \tfrac{1}{4}[D(y_0) + D(y_i) + D(y_j) + D(y_k)] = \sigma^2; \\ &(i = 1, 2, 3). \end{aligned}$$

It can be concluded that the "rational" design provides better results than the "standard" design.

We can object, however, that according to the "standard" design, we are weighing each object separately, that means independently. This independence is only illusory. Indeed, computing the covariance of $\hat{\alpha}_i$ and $\hat{\alpha}_j$ in the "standard" case, we obtain from Eq. (1)

$$\begin{aligned} \operatorname{cov}(\hat{\alpha}_i, \hat{\alpha}_j) = E\{&[(y_i - \alpha_i) - (y_0 - \alpha_0)] \\ \times &[(y_j - \alpha_j) - (y_0 - \alpha_0)]\} = D(y_0) = \sigma^2 \neq 0. \end{aligned}$$

On the other hand, for the estimates given in Eq. (7) we obtain

$$\operatorname{cov}(\hat{\alpha}_i, \hat{\alpha}_j) = 0; \quad (i \neq j).$$

Thus the independence appears just in the "rational" case.

EXAMPLE 2. *Measurement of the power of a car.*
The power of the engine of a car is tested on a testing road during the accelerated motion of the car. We can state, after a simplification, that the car is accelerated uniformly on the test segment, according to the equation

$$s(t) = vt + \tfrac{1}{2}zt^2. \tag{8}$$

Here $s(t)$ is the position of the car at time t (supposing that $s(0)=0$), v is the velocity at $t=0$, and z is the acceleration which is constant during the test. The parameters v and z, being unknown, are to be estimated from the observed data. The major purpose of the experiment is to specify the value of the parameter z which indicates the power of the engine.

The used measuring devices can perform 10 independent measurements of the positions of the car at times $t_1, \ldots, t_{10}$ chosen arbitrarily from the time interval 0—10 s. Simultaneous measurements at the same time (i.e. $t_i = t_j$) are allowed.

The optimization is to be done in two stages. First, formulae for the best estimates of v and z are to be established, and then, based on those formulae, the best times $t_1, \ldots, t_{10}$ are to be chosen.

The observed measured quantities are random variables $y(t_1), \ldots, y(t_{10})$ which have means and variances equal to

$$\begin{aligned} &\mathrm{E}[y(t_i)] = vt_i + z\frac{t_i^2}{2}, \\ &\mathrm{D}[y(t_i)] = \sigma^2; \quad (i=1, \ldots, 10). \end{aligned} \tag{9}$$

The variance σ^2 is unknown, which has no influence on the parameter estimation and on the design, as will be seen later. The estimates $\hat{v}$, $\hat{z}$ are to be estimated from the data $y(t_1), \ldots, y(t_{10})$, respecting that

$$\mathrm{E}(\hat{v}) = v, \quad \mathrm{E}(\hat{z}) = z$$

and that the variances $\mathrm{D}(\hat{v})$ and $\mathrm{D}(\hat{z})$ are to be minimum. In accordance with the notation in Chapter II, let us denote by $\boldsymbol{f}(t)$ the vector

$$\boldsymbol{f}(t) = \begin{pmatrix} t \\ t^2/2 \end{pmatrix}$$

The information matrix of the design $t_1, \ldots, t_{10}$ is equal to (cf. Eq. (4) in Chapter II)

$$\mathbf{M} = \sum_{k=1}^{10} \boldsymbol{f}(t_k)\boldsymbol{f}'(t_k)\sigma^{-2} = \frac{1}{\sigma^2}\begin{pmatrix} \sum_{k=1}^{10} t_k^2, & \sum_{k=1}^{10} t_k^3/2 \\ \sum_{k=1}^{10} t_k^3/2, & \sum_{k=1}^{10} t_k^4/4 \end{pmatrix}. \tag{10}$$

The estimate of z, that is the best under the design $t_1, \ldots, t_{10}$, is equal to

$$\hat{z} = (0, 1)\mathbf{M}^{-} \sum_{k=1}^{10} \boldsymbol{f}(t_k) y(t_k) \sigma^{-2} \tag{11}$$

and it does not depend on σ^2. Here the matrix $\mathbf{M}^-$ is an arbitrary solution of the equation

$$\mathbf{M} = \mathbf{M}\mathbf{M}^-\mathbf{M}$$

(cf. Eq. (13) in Chapter II). Especially, if $\det \mathbf{M} \neq 0$, then $\mathbf{M}^- = \mathbf{M}^{-1}$.
The variance of the estimate z is equal to

$$D(\hat{z}) = (0, 1)\mathbf{M}^- \begin{pmatrix} 0 \\ 1 \end{pmatrix} = \{\mathbf{M}^-\}_{22}. \tag{12}$$

The design yielding a minimum value of $D(\hat{z})$ can be obtained using a graphical method based on Proposition III.17. The curve in Fig. 1 is the set

$$T \equiv \{\boldsymbol{f}(t) : t \in \langle 0, 10 \rangle\} \cup \{-f(t) : t \in \langle 0, 10 \rangle\}.$$

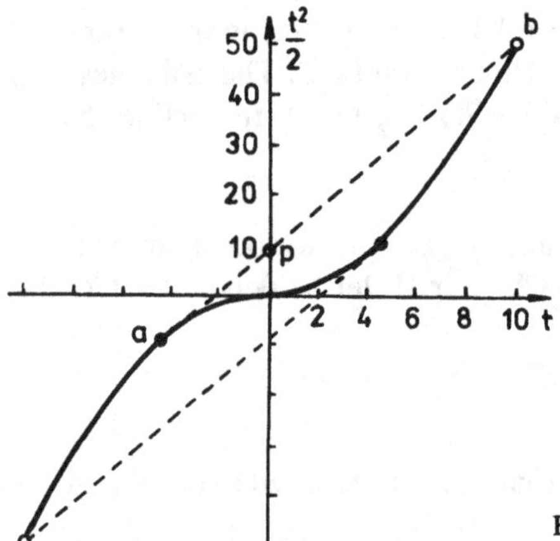

Fig. 1.

The dashed line marks the boundary of the convex hull S of the set T. Let us denote by $\boldsymbol{p}$ one of the points of intersection of the boundary of S with the straight line passing through the origin in the direction of the vector $\boldsymbol{g} \equiv (0, 1)'$. Let $\boldsymbol{a} \equiv (-t, -t^2/2)$ and $\boldsymbol{b} \equiv (t^*, t^{*2}/2)$ be two points belonging to the set T, such that $\boldsymbol{p}$ is on the abscissa connecting $\boldsymbol{a}$ with $\boldsymbol{b}$. From Fig. 1 it can be seen that

$$t = 4.4 \text{ s}, \quad t^* = 10 \text{ s}$$

and that the components of the vector $\boldsymbol{p}$ are

$$\begin{pmatrix} p_1 \\ p_2 \end{pmatrix} = \xi_1 \begin{pmatrix} -t \\ -t^2/2 \end{pmatrix} + \xi_2 \begin{pmatrix} t^* \\ t^{*2}/2 \end{pmatrix},$$

where

$$\xi_1 = \|\boldsymbol{b} - \boldsymbol{p}\| / \|\boldsymbol{a} - \boldsymbol{b}\| = 0.7,$$
$$\xi_2 = \|\boldsymbol{a} - \boldsymbol{p}\| / \|\boldsymbol{a} - \boldsymbol{b}\| = 0.3.$$

According to Proposition III.17, an optimum desing is to be allocated at the points t and t^*, the number of independent observations being proportional to ξ_1 and ξ_2. That means that the optimum design is

$$t_1 = \ldots = t_7 = 4.4 \text{ s},$$
$$t_8 = \ldots = t_{10} = 10 \text{ s}.$$

From Fig. 1 and Proposition III.17 it follows that in the case of an optimum design

$$\mathrm{D}(\hat{z}) \doteq \sigma^2/9.$$

Sometimes it is more convenient to repeat a few trials independently than to do many different trials in the same experiment. This is illustrated by the following example.

EXAMPLE 3 [4]. Consider observing the values of a second order polynomial

$$\vartheta(x) = \alpha_1 + \alpha_2 x + \alpha_3 x^2 \tag{13}$$

at some points of the interval $\langle 0, 1 \rangle$. The observations are independent having random errors with a constant variance σ^2. The parameters α_1, α_2 are unknown. The purpose of the experiment is to estimate the value of

$$h(\alpha_1, \alpha_2, \alpha_3) \equiv \alpha_1 - \alpha_2 + \alpha_3. \tag{14}$$

It can be checked, using the Corollary to Proposition VI.8, that the optimum design requires 35% of observations at the point $x^{(1)}=0$, 47% at the point $x^{(2)}=0.5\%$, and 18% at the point $x^{(3)}=1$. Let D_{opt} be the variance of the best linear estimate for $h(\alpha_1, \alpha_2, \alpha_3)$, under the optimum design. Let D_n be the variance of the best linear estimate for $h(\alpha_1, \alpha_2, \alpha_3)$, the observations being allocated uniformly at n points

$$x^{(i)}=(i-1)/(n-1); \quad (i=1, \ldots, n).$$

When n tends to infinity, D_n tends to a value about 3 times larger than D_{opt} (see Fig. 2).

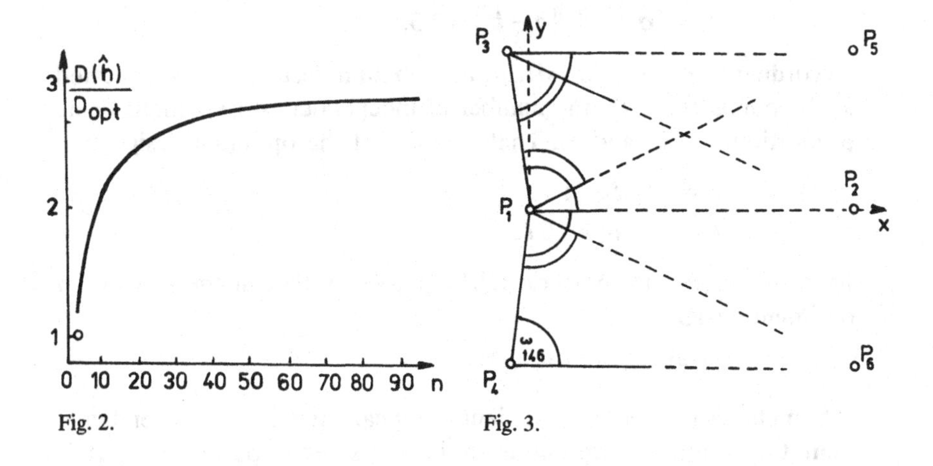

Fig. 2. Fig. 3.

The mean values of the random variables observed in the Examples 1—3 are linear functions of the parameters $\alpha_1, \alpha_2, \ldots$ [Eqs. (3), (6), (9), (13)]. In many real experiments similar functions are nonlinear. To use the results of this book we must linearize the nonlinear functions, and eventually evaluate the bias arising in the estimates due to linearization. For details see Chapter II.4, Proposition II.10.

Example 4. *The estimation of the length of a bridge* [22].
Consider an experiment to measure the length of a planned bridge across the river (the distance of the points P_1 and P_2 in Fig. 3). It is not

possible to make any measurements above water surface using precise optical distance measuring instruments, because of the light reflections disturbing the needed accuracy. Therefore it is necessary to measure the length of the bridge indirectly, from 3 points P_1, P_3, P_4, on one side of the river, and from P_2, P_5, P_6 on the other side (Fig. 3). In the whole geodetic net in Fig. 3 it is allowed to measure the distances d_{ij} between any two points P_i, P_j on the same side of the river and the angles $\omega_{ijk} = \sphericalangle P_iP_jP_k$ defined by any 3 points P_i, P_j, P_k. The length of the bridge is computed indirectly, using trigonometric formulae. It is necessary to repeat many times the measurements of the chosen distances and angles, and to use statistics to ensure the needed precision. Hence an optimum allocation of measurements of distances and angles becomes necessary.

It is allowed to measure 6 different distances and 60 different angles. The variance of d_{13}, d_{14}, d_{25} or d_{26} is equal to $(3 \times 10^{-3}\ \mathrm{m})^2$. The variance of d_{34} or d_{56} is $2 \times (3 \times 10^{-3}\ \mathrm{m})^2$, the variance of each angle is $(2.062'')^2$.

Take the length of the bridge (= the parameter α_1) and the coordinates of the points $P_3 = (\alpha_2, \alpha_3)$, $P_4 = (\alpha_4, \alpha_5)$, $P_5 = (\alpha_6, \alpha_7)$, $P_6 = (\alpha_8, \alpha_9)$ as independent parameters. The coordinate axes are oriented so that $P_1 = (0, 0)$ and $P_2 = (\alpha_1, 0)$. The relations between these parameters and the distances and angles are

$$
\begin{aligned}
d_{13} &= [\alpha_2^2 + \alpha_3^2]^{1/2},\\
d_{14} &= [\alpha_4^2 + \alpha_5^2]^{1/2},\\
&\cdots\cdots\cdots\cdots\cdots\cdots\\
\omega_{123} &= \mathrm{arctg}\,[\alpha_3/(\alpha_1 - \alpha_2)],\\
\omega_{215} &= \mathrm{arctg}\,(\alpha_7/\alpha_6).\\
&\cdots\cdots\cdots\cdots\cdots\cdots
\end{aligned}
\qquad (15)
$$

The functions on the right-hand side of Eqs. (15) are nonlinear. They can be linearized, i.e. approximated by the linear term in the Taylor expansion in the neighbourhood of a point $\alpha_1^0, \ldots, \alpha_9^0$, because of small variances of the errors. The values of $\alpha_1^0, \ldots, \alpha_9^0$ can be obtained from simple measurements of the coordinates of the points $P_1, \ldots, P_6$.

The optimum design cannot be obtained by the method used in Example 2, the number of parameters $\alpha_1, \ldots, \alpha_9$ being too large for that. However, an almost optimum design can be obtained using some of the iterative methods that are described in Chapter V. A proper method is given in Chapter V.2. According to it, the computation is started by choosing an arbitrary design that allows estimating all parameters $\alpha_1, \ldots, \alpha_9$. This design is corrected step by step (i.e. iteratively), till a simple stopping rule indicates that an "almost optimum" design has been attained. This iterative method can be used effectively on computers only.

The resulting design computed in this way is (cf. [22]): the lengths d_{12} (34% of measurements), d_{14} (24%), and the angles ω_{132} (25%), ω_{324} (29%), ω_{432} (9%) and ω_{241} (1%) are to be measured.

The model of any experiment in the examples is the regression model. The formal definition of the regression model is simple (Definition II.1) and it includes different kinds of experiments: with direct or indirect observation, factorial models, variance — analysis models, etc. (cf. Chapter II.2). Estimation in this model is closely connected with the properties of the information matrices of designs (Definitions II.4 and II.5). It is possible to estimate without using information matrices, as mentioned in some exercises and as follows from Chapter VII. However, this is not advantageous since information matrices make the numerical solution of estimation and design problems much easier.

An experiment is composed of different trials, hence the set of all trials possible in a particular experiment is to be specified first. It appears to be useful to consider any probability distribution supported by a finite set of trials as an experimental design in an asymptotical sense (cf. Definition II.3). Then the set of all information matrices, corresponding to all possible designs in a given experiment, is convex. Functions defining the optimality criteria appear to be convex as well. So, properties of convex functions are exploited in large part of the book.

Alternative estimates to the best linear unbiassed estimates are considered in Chapter II.4: the nonlinear estimates in a Gaussian regression model, robust estimates and ridge estimates.

Some deviations from the used regression model are discussed in Chapter II.5. They are: nonlinearity, statistical dependence between the observed variables, the change of the dimension of the model (= of the number of unknown parameters).

A uniform ordering of designs is considered in Chapter III. A design is considered to be uniformly better than another design if it guarantees a smaller variance of estimates of any linear function of the unknown parameters. This uniform ordering of designs can be expressed by an ordering of information matrices, as shown in Propositions III.1—III.3. Since, in general, there are no uniformly best designs, admissible designs are to be considered. It is proved in Proposition III.7 that the restriction to admissible designs may lead to an essential reduction of the set of trials which are to be considered.

A useful refinement of the uniform ordering of designs is considered in Propositions III.4 and III.5.

The variance of an estimate, considered as a function of the information matrix, is analysed in Chapter III (Propositions III.13 and III.14). This function can have discontinuities which can yield to difficulties when computing optimum designs.

In Chapter IV optimality criteria are discussed. They are used to express the aim of the experiment mathematically. A large amount of different criteria is to be considered to give a greater flexibility when expressing this aim. An optimality criterion is usually represented by a function ϕ defined on the set of all information matrices. Analytical properties of such functions as continuity, convexity, differentiability are discussed.

Optimality criteria are classified in two groups: global criteria, when all unknown parameters in the model are considered important; and partial criteria, when only some parameters or some functions of the parameters are important.

Convexity is an essential property of all considered criteria functions. Therefore, consequences of convexity are discussed at the end of Chapter IV. The most important is the so-called "equivalence theorem" (Propositions IV.26 and IV.27) that states simple sufficient and necessary conditions of the optimality of a design. The "almost optimality" can be checked by similar tools, as presented in Proposition IV.28.

Different methods for computing optimum experimental designs are presented in the book. A simple method of establishing the design minimizing the variance of an estimate is given in Proposition III.17. This method can be used graphically, if the number of unknown parameters in the model is not larger than 3. Several direct methods of computing optimum designs in special cases are considered in Chapter VI. However, the main stress is on iterative methods, which are discussed in Chapter V. Two iterative methods for computing the D-optimum designs are presented in detail, other methods for D-optimality are explained only briefly. A general approach to the iterative procedures is presented in Chapter V.4. Its purpose is to give a certain theoretical basis to the reader to prove the convergence of his own iterative method. The use of that is illustrated in Chapter V.6.4 where a method is presented for computing designs which are D-optimum in the class of all designs making it possible to test the significance of an additional parameter in the model, on a given significance level. A survey of some published iterative methods is presented in Chapters V.5 and V.6.

Iterative methods are largely universal but do not allow one to gain insight into the structure of optimum designs. Therefore, in Chapter VI the optimum designs are established analytically for special cases, using the methods of approximation of functions or by a direct investigation of the properties of information matrices. In such a way optimum designs can be computed for the "first order regression model", for the polynomial regression on a one-dimensional interval, and for the trigonometric regression. Notice that Chapter VI can be read without reading Chapter V.

In Chapter VII instead of the regression model another model is discussed. It is a generalization of the regression model to the case of an infinite number of unknown parameters (i.e. to a model with infinitely many "degrees of freedom"). This model can be motivated by a study of physical fields. Functional analysis instead of matrix algebra is the main mathematical tool in this model. Because the regression model can be considered as a special case of this more general model, Chapter VII presents a different approach to the results in Chapters II—IV. In Chapter VII.3 the estimates of linear functionals are considered. It is shown that the estimability of linear functionals is closely connected with their continuity (Propositions VII.2 and VII.3). As a consequence, the design which is optimum for the estimation of a given functional can be obtained from the Riesz integral representation of the functional (Proposition VII.3). In the last part of Chapter VII the properties of the so-called Wiener chaos are used to establish the properties of the estimates of polynomials. As a consequence, a new optimality criterion is obtained.

Now we shall give some *notes about the procedure which can be recommended when an experiment is to be designed optimally.*

The procedure has three stages:

A) Specification of the model of the experiment.
B) Computation of the optimum design.
C) Analysis of the optimum design.

A) *To specify the model it is advisable*:

a) to determine the trials which can be performed in the experiment (i.e. to define the set X, see Chapter II.2);

b) to specify the relationships between the unknown parameters $\alpha_1, \ldots, \alpha_m$ of the observed object and between the "theoretical values" (i.e. the mean values) of the observed variables $y(x)$; $(x \in X)$ (see Chapter II.2);

c) if the obtained relationships are nonlinear, to try to linearize them and to evaluate the error (= bias) which follows from the linearization;

d) to specify (up to a multiplicative constant) the accuracy of the measurements in the trial, i.e. to determiny the variance $\sigma^2(x) = D(y(x))$;

e) if the obtained linear (or linearized) relationships (points b and c) are not of the standard form that is considered in Chapter II.2, model (b), a change of the parameters considered in the model is necessary (see the different kinds of the regression models (a)—(h) in Chapter II.2, which can be reparametrized to obtain the model (b));

f) it can be useful to exclude a priori some trials that do not give much information. For this purpose use Proposition III.6;

g) it is useful to specify the purpose of the experiment as precisely as possible, and to choose an optimality criterion expressing this purpose. If this is not possible, try to specify a new optimality criterion of your own. Eventually modify the obtained criterion function till obtaining a convex function (e.g. by a monotone transformation of the criterion function). Determine other properties of the criterion function (differentiability).

B) *To compute the optimum design it can be recommended*:

a) to check whether the considered experiment does correspond to some of the special cases discussed in Chapter VI. If so, use the methods and results presented in this chapter;

b) to consider some "reasonable" designs chosen a priori. Use the rules given in Propositions IV.27 and IV.28 to check "how far" are the considered designs from the optimum designs;

c) if the results obtained according to a) and b) are not satisfactory, it can be recommended to use some of the iterative methods presented in Chapter V and to compute the optimum design on a computer;

d) if there is no iterative method corresponding to your optimality criterion then you have to prepare such a method using the theory presented in Chapter V.5.

C) *The analysis of a computed design consists of*:

a) checking the optimality of the design with respect to other alternative optimality criteria (using Proposition IV.28);

b) making some changes in the computed design to get some "useful" properties (e.g. to diminish the size of the set of trials involved in the design by grouping "neighbourhood" trials, to find a fixed size design which is close to the computed design, to make a symmetrization of the design, etc.). Each change must be checked using Proposition IV.28.

Chapter II

The Regression Model and Methods of Estimation

II.1. INTRODUCTION

Mathematical methods are advantageously used in formulating the model of an experiment, in analysing the data obtained from an experiment and in designing an experiment. This book deals with experimental design but an introductory consideration of modelling and estimation are expedient. They are presented in this chapter and in the beginning of Chapter VII. An extensive exposition of linear estimation can be found in [88], a standard use of methods is explained, e.g., in [95].

The structure of the regression model considered in this book is essentially linear. Therefore linear estimates are considered here mainly, nonlinear estimates are discussed in Chapters II.4 and VII.

This chapter contains the definition and interpretation of a linear regression model with uncorrelated observations (Definition II.1; the model with correlated observations is mentioned in Chapter II.4). An asymptotical approach to the definition of a design (see Definition II.3) gives the possibility to use methods of convex analysis for designing experiments, as explained in the next chapters.

The central concept in the linear estimation is the Fisher information matrix (see Definition II.4). The calculus of information matrices is a useful tool for computing the best linear estimates or for designing an experiment. The less known expression for the variance of the best linear estimate of a linear functional which is given in Eq. (23), is equal to the norm of this functional in a certain (finite-dimensional) Hilbert space. This indicates the connection with Hilbert space methods, but only in the infinite-dimensional case, in Chapter VII such methods are to be used instead of the information matrices.

The final part of the chapter centres upon deviations from the used methods of estimation and from the regression model with uncorrelated observations.

II.2. THE REGRESSION MODEL WITH UNCORRELATED OBSERVATIONS

The experimenter preparing an experiment disposes of a certain amount of possible partial, elementary experiments which can be used as components of the whole designed experiment. Such partial experiments will be called trials. Denote the set of all trials which can be performed in an experiment by X. The result of an observation in the trial $x \in X$ (= in the "point" $x \in X$) is the value of a real random variable $y(x)$ having a known variance

$$\mathrm{E}[y(x)-\mathrm{E}(y(x))]^2$$

and an unknown mean $\mathrm{E}[y(x)]$. Note that the case when several variables can be observed in one trial is mentioned in Chapter II.5.

A state is described in the regression model by a function

$$\vartheta : x \in X \mapsto \mathrm{E}[y(x)] \in R.$$

We shall write $\mathrm{E}_\vartheta[y(x)]$ instead of $\mathrm{E}[y(x)]$ to stress the dependence of the mean value on the state ϑ.

DEFINITION II.1. *The linear regression model with uncorrelated observation* (*or briefly*: *the regression model*) *is a triplet* (X, Θ, σ), *where* X *is a compact metric space*, Θ *is a finite-dimensional linear space of functions that are defined and continuous on* X, *and* σ *is a continuous positive function defined on* X.

The set X is *the set of trials*, Θ is *the set of states* of the observed object, σ^{-1} is *the function of precision* of observations. Uncorrelated random variables $y(x)$; $(x \in X)$ can be observed. They are such that

$$\mathrm{E}_\vartheta[y(x)] = \vartheta(x),$$
$$\mathrm{D}[y(x)] = \mathrm{E}_\vartheta[y(x)-\mathrm{E}_\vartheta(y(x))]^2 = \sigma^2(x); \quad (x \in X, \vartheta \in \Theta). \quad (1)$$

REMARKS. (1) Before performing an experiment only the sets X, Θ and the function σ are known. The unique information about the state is that $\vartheta \in \Theta$. The function σ is usually specified by the precision of the used measuring instruments. To design an experiment it is sufficient to known the function σ up to a multiplicative constant.

(2) The function $\vartheta \in \Theta$ (*the state* of the object) is sometimes called *the response function*. This is a term taken over from biological experiments, $\vartheta(x)$ being the response to the stimulus x.

(3) The generally accepted assumptions that X is a compact metric space, and that ϑ and σ are continuous, are not used in this chapter. They are exploited later. The assumptions on X are satisfied in the important cases when either X is finite, or

$$X = \langle a_1, b_1 \rangle \times \langle a_2, b_2 \rangle \times \ldots \times \langle a_k, b_k \rangle$$

is a closed k-dimensional interval. In the second case this means that every trial $x \in X$ is specified by the levels of k controlled factors, the i-th of them being bounded by a_i and b_i.

DEFINITION II.2. *A fixed size design (or an "exact" design) with the size N is a sequence of N trials $x_1, \ldots, x_N$.*

REMARK. When the experiment is designed according to a fixed size design $x_1, \ldots, x_N$, uncorrelated observations of the random variables $y(x_1), \ldots, y(x_N)$ are performed. Some of the points $x_1, \ldots, x_N$ may coincide; it means that observations are repeated independently in the same trial.

Denote the number of repetitions of the trial x in the sequence $x_1, \ldots, x_N$ by N_x.

DEFINITION II.3. *The discrete measure ξ defined by*

$$\xi(x) = \frac{N_x}{N}; \quad (x \in X) \tag{2}$$

is the design associated with the fixed-size design $x_1, \ldots, x_N$. A discrete probability measure on X which is supported by a finite set $X_\xi \equiv \{x \colon x \in X, \xi(x) > 0\}$ is an (asymptotical) design.

REMARK. Let ξ be a design and N a positive integer. The numbers

$N_x \equiv N\xi(x)$; $(x \in X)$ may not be integers. Hence, many designs are associated with no design of size N. Nevertheless, with N tending to infinity they are approximating fixed-size designs very well.

It is usual to distinguish between different kinds of regression models, according to the specification of the set Θ. Let us present some of them:

a) *The model of direct observations*. In this model there is only one unknown parameter α which is measured directly in every trial. The set of states is

$$\Theta = \{\alpha j : \alpha \in R\},$$

where $j : x \in X \mapsto 1 \in R$. That means that the trials differ only by the value of $D[y(x)] = \sigma^2(x)$.

b) *The model of indirect observations*. Let $f_1, \ldots, f_m$ be given continuous and linearly independent functions defined on X. Let

$$\Theta = \left\{ \sum_{i=1}^{m} \alpha_i f_i : (\alpha_1, \ldots, \alpha_m)' \in R^m \right\}.$$

In this model m parameters $\alpha_1, \ldots, \alpha_m$ are estimated "indirectly", based on the observed values of $y(x_1), \ldots, y(x_N)$. Evidently

$$E_\vartheta[y(x)] = \sum_{i=1}^{m} \alpha_i f_i(x); \quad \left(x \in X, \ \vartheta = \sum_{i=1}^{m} \alpha_i f_i \right).$$

This model is central in further considerations since any regression model can be reduced to the model (b) choosing an arbitrary linear basis $f_1, \ldots, f_m$ in Θ.

c) *The model with linear couplings*. Let X be a finite set, and let $b_1, \ldots, b_q$ be linearly independent functions defined on X. Define

$$\Theta = \left\{ \vartheta : \vartheta \in C(X), \sum_{x \in X} b_i(x) \vartheta(x) = 0; \quad (i = 1, \ldots, q) \right\}.$$

d) *The model of indirect observations under linear constrains*. Let the range of the parameters $\alpha_1, \ldots, \alpha_m$ in the model (b) be restricted by linear constrains

$$\sum_{j=1}^{m} B_{lj} \alpha_j = 0; \quad (l = 1, \ldots, q),$$

where B_{lj}; $(l=1, \ldots, q, j=1, \ldots, m)$ are given coefficients. That means

$$\Theta=\left\{\vartheta: \vartheta=\sum_{i=1}^{m} \alpha_i f_i, \sum_{j=1}^{m} B_{lj}\alpha_j=0,\right.$$

$$\left.(\alpha_1, \ldots, \alpha_m)' \in R^m, (l=1, \ldots, q)\right\}.$$

e) *The model of the "spline regression"*. Take $k+1$ points $x_0=a< x_1<\ldots<x_k=b$ from the interval $X=\langle a, b\rangle$. Let $f_1^{(i)}, \ldots, f_m^{(i)}$ be functions which are linearly independent on $\langle x_{i-1}, x_i\rangle$ and r-times continuously differentiable in the open interval (x_{i-1}, x_i). Let Θ be the set of all functions $\vartheta \in C(\langle a, b\rangle)$ such that

1. ϑ restricted to $\langle x_{i-1}, x_i\rangle$ is in the span of the set $\{f_1^{(i)}, \ldots, f_m^{(i)}\}$.

2. $$\lim_{x \searrow x_i} \frac{d^j\vartheta(x)}{dx^j}=\lim_{x \nearrow x_i} \frac{d^j\vartheta(x)}{dx^j}; \quad (i=1, \ldots, k-1,\ j=0, \ldots, r).$$

f) *The model of the two-level factorial experiment*. Let

$$X=\{(x_1, \ldots, x_k): x_i \in \{-1, 1\};\ i=1, \ldots, k\}.$$

The numbers -1 and 1 are the possible levels of any factor x_i. Let

$$\Theta=\left\{\vartheta: \vartheta(x)=\alpha_0+\sum_{i=1}^{k} \alpha_i x_i+\sum_{1 \leq i<j \leq k} \alpha_{i,j} x_i x_j,\right.$$

$$\left.\alpha_0, \ldots, \alpha_k, \alpha_{1,2}, \ldots, \alpha_{k-1,k} \in R^{1+k(k+1)/2}\right\}.$$

More complicated three- and more-level factorial models can be considered (cf. [10]).

g) *The block experiment in the analysis of variance*. Let

$$X=\{(i, j): i \in\{1, \ldots, v\}, j \in\{1, \ldots, b\}\}.$$

The level i is interpreted as the i-th treatment, and the level j is interpreted as the j-th block. The model under consideration is given by

$$E[y(i, j)]=\mu+\alpha_i+\beta_j,$$

where μ is said to be the "mean effect", α_i — the "effect of the i-th treatment", and β_j — the "effect of the j-th block". The parameters α_i and β_j satisfy the constrains

$$\sum_{i=1}^{v} \alpha_i = \sum_{j=1}^{b} \beta_j = 0.$$

Hence

$$\Theta = \left\{ \vartheta : \vartheta(i, j) = \mu + \alpha_i + \beta_j, \sum_{i=1}^{v} \alpha_i = 0, \sum_{j=1}^{b} \beta_j = 0, \right.$$
$$\left. (\mu, \alpha_1, \ldots, \alpha_v, \beta_1, \ldots, \beta_b) \in R^{v+b+1} \right\}.$$

More complicated block experiments can be considered [35, 37, 38].

h) *The model defined by a symmetry requirement*. Let G be a group of one-to-one mappings of X onto X, expressing the symmetry of the set X. The group product in G is the composition of mappings. Suppose that G has a finite amount of orbits. Note that an orbit is a maximal set $\tilde{X} \subset X$ such that $x_1, x_2 \in X$ implies that there is a mapping $g \in G$ such that $x_2 = g(x_1)$. The set of states in the model is given by

$$\Theta = \{\vartheta \in C(X) : \vartheta[g(x)] = \vartheta(x);\ (g \in G, x \in X)\}.$$

II.3. ESTIMATES OF LINEAR FUNCTIONALS IN THE REGRESSION MODEL

Functionals defined on the set of states Θ (i.e. real-valued functions defined on the linear space Θ) can be interpreted as variables (i.e. the physical quantities) characterizing the observed object. The main attention will be paid to linear functionals, in accordance with the linear structure of the regression model.

A linear (homogeneous) functional g is a real-valued function defined on Θ such that $g(\alpha\vartheta_1 + \beta\vartheta_2) = \alpha g(\vartheta_1) + \beta g(\vartheta_2)$ for every ϑ_1, $\vartheta_2 \in \Theta$, $\alpha, \beta \in R$.

EXAMPLES. 1. Take $x \in X$, and define

$$g_x : \vartheta \in \Theta \mapsto \vartheta(x) \in R.$$

2. The functional

$$g_j : \vartheta = \sum_{i=1}^{m} \alpha_i f_i \in \Theta \mapsto \alpha_j \in R$$

defines the parameter α_j in the model of indirect observations.

3. In the same model let us define

$$g: \vartheta = \sum_{i=1}^{m} \alpha_i f_i \in \Theta \mapsto \sum_{i=1}^{m} g_i \alpha_i \in R,$$

where $g_1, \ldots, g_m$ are given coefficients. Notice that any linear functional can be expressed in this way.

Consider a fixed size design $x_1, \ldots, x_N$ in the regression model (X, Θ, σ), and a linear functional g on Θ. The value $g(\vartheta)$ is unknown since the true state ϑ is unknown. This value is to be estimated from the experimental data by a linear function of the random variables $y(x_1)$, $\ldots$, $y(x_N)$,

$$\sum_{i=1}^{N} c_i y(x_i),$$

where $c_1, \ldots, c_N$ are some coefficients. This function or its value is said to be a *linear estimate* of g. The estimate is *unbiassed* iff

$$\mathrm{E}_\vartheta \left[\sum_{i=1}^{N} c_i y(x_i) \right] = g(\vartheta); \quad (\vartheta \in \Theta).$$

The functional g is said to be *estimable* if there is at least one unbiassed estimate of g. *The best linear estimate* (or the BLUE) for g is the unbiassed estimate having the minimum variance.

Observe that nonlinear estimates of g are considered in Proposition II.8.

PROPOSITION II.1. *If $c_1^*, \ldots, c_N^*$ is the solution of the minimization problem*

$$\min \left\{ \sum_{i=1}^{N} c_i^2 \sigma_i^2 : (c_1, \ldots, c_N)' \in R^N, \sum_{i=1}^{N} c_i \vartheta(x_i) = g(\vartheta);\ \vartheta \in \Theta \right\},$$

then

$$\sum_{i=1}^{N} c_i^* y(x_i)$$

is the BLUE for g.

The proof of this statement is evident since $\sum_{i=1}^{N} c_i^2\sigma_i^2$ is the variance of $\sum_{i=1}^{N} c_i y(x_i)$. □

Later we shall obtain an explicit expression for $c_1^*, \ldots, c_N^*$ using matrix technique.

To proceed further let's take linearly independent functions $f_1, \ldots, f_m \in \Theta$ which constitute a linear basis of Θ. That means $f_1, \ldots, f_m$ are such that to every $\vartheta \in \Theta$ there is a unique vector $\boldsymbol{\alpha} = (\alpha_1, \ldots, \alpha_m)$ such that

$$\vartheta(x) = \sum_{i=1}^{m} \alpha_i f_i(x); \quad (x \in X).$$

Similarly, to every linear functional g on Θ there is a unique vector $\mathbf{g} = (g_1, \ldots, g_m)'$ such that

$$g(\vartheta) = \sum_{i=1}^{m} g_i \alpha_i; \quad \left(\vartheta = \sum_{i=1}^{m} \alpha_i f_i \in \Theta\right).$$

Let $\mathbf{y} \equiv (y(x_1), \ldots, y(x_N))'$ be the vector of observed random variables, let $\mathbf{c} \equiv (c_1, \ldots, c_N)'$ and $\mathbf{f}(x) \equiv (f_1(x), \ldots, f_m(x))'$ be vectors of coefficients. Denote by $\mathbf{F}$ the $N \times m$ matrix

$$\mathbf{F} = \begin{pmatrix} \mathbf{f}'(x_1) \\ \vdots \\ \mathbf{f}'(x_N) \end{pmatrix} = \begin{pmatrix} f_1(x_1), \ldots, f_m(x_1) \\ \vdots \\ f_1(x_N), \ldots, f_m(x_N) \end{pmatrix}. \tag{3}$$

DEFINITION II.4. *The information matrix of the fixed-size design* $x_1, \ldots, x_N$ *is the matrix*

$$\mathbf{M} = \sum_{i=1}^{N} \mathbf{f}(x_i)\mathbf{f}'(x_i)\sigma^{-2}(x_i). \tag{4}$$

NOTE: A symmetric $m \times m$ matrix $\mathbf{A}$ is *positive semidefinite*, denote

$$\mathbf{A} \geqslant \mathbf{0}$$

iff $\mathbf{u}'\mathbf{A}\mathbf{u} \geqslant 0$ for every $\mathbf{u} \in R^m$. It is *positive definite* iff moreover $\mathbf{u}'\mathbf{A}\mathbf{u} = 0$ implies $\mathbf{u} = \mathbf{0}$. Denote

$$\mathbf{A} > \mathbf{0}.$$

Evidently, the information matrix $\mathbf{M}$ is positive semidefinite. It is positive definite if and only if $\det \mathbf{M} \neq 0$.

PROPOSITION II.2. (*The fundamental proposition about linear estimates.*)
Let $x_1, \ldots, x_N$ *be a fixed-size design in the regression model* (X, Θ, σ) *and let* g *be a linear functional defined on* Θ.
The functional g *is estimable if and only if the vector* $\mathbf{g}$ *(corresponding to* g*) is a linear combination of the columns of the matrix* $\mathbf{M}$, *i.e.* $g \in \mathcal{M}(\mathbf{M})$, *where*

$$\mathcal{M}(\mathbf{M}) = \{\mathbf{M}\boldsymbol{u} : \boldsymbol{u} \in R^m\}. \tag{5}$$

In such a case there is a unique vector $\mathbf{z}_g \in \mathcal{M}(\mathbf{M})$ *such that* $\mathbf{g} = \mathbf{M}\mathbf{z}_g$. *The BLUE for* g *is*

$$\hat{g} = \sum_{i=1}^{N} \mathbf{z}_g' \boldsymbol{f}(x_i) \sigma^{-2}(x_i) y(x_i). \tag{6}$$

The variance of g *is equal to*

$$D(\hat{g}) = \mathbf{z}_g' \mathbf{M} \mathbf{z}_g = \sup \left\{ \frac{(\mathbf{g}'\boldsymbol{\alpha})^2}{\boldsymbol{\alpha}'\mathbf{M}\boldsymbol{\alpha}} : \boldsymbol{\alpha} \in R^m, \mathbf{M}\boldsymbol{\alpha} \neq \mathbf{0} \right\}. \tag{7}$$

PROPOSITION II.3. *There is* $\boldsymbol{u} \in R^m$ *such that* $\mathbf{g} = \mathbf{M}\boldsymbol{u}$ *if and only if there is* $\boldsymbol{c} \in R^N$ *such that* $\mathbf{g} = \mathbf{F}'\boldsymbol{c}$, *where* $\mathbf{F}$ *is the matrix defined by Eq.* (3).

PROOF. Let $\boldsymbol{\Sigma}$ be the diagonal matrix, $\boldsymbol{\Sigma} \equiv \mathrm{diag}\,(\sigma(x_1), \ldots, \sigma(x_N))$. The information matrix can be expressed as

$$\mathbf{M} = \mathbf{F}'\boldsymbol{\Sigma}^{-2}\mathbf{F}.$$

From $\mathbf{g} = \mathbf{M}\boldsymbol{u} = \mathbf{F}'\boldsymbol{\Sigma}^{-2}\mathbf{F}\boldsymbol{u}$ we obtain $\mathbf{g} = \mathbf{F}'\boldsymbol{c}$ taking $\boldsymbol{c} = \boldsymbol{\Sigma}^{-2}\mathbf{F}\boldsymbol{u}$. Conversely, from $\mathbf{g} = \mathbf{F}'\boldsymbol{c}$ it follows that $\mathbf{g} = (\boldsymbol{\Sigma}^{-1}\mathbf{F})'\boldsymbol{\Sigma}\boldsymbol{c}$. Let $\boldsymbol{w}$ be the vector obtained by the orthogonal projection of $\boldsymbol{\Sigma}\boldsymbol{c}$ onto $\mathcal{M}(\boldsymbol{\Sigma}^{-1}\mathbf{F})$. Hence

$$(\boldsymbol{\Sigma}^{-1}\mathbf{F})'\boldsymbol{w} = (\boldsymbol{\Sigma}^{-1}\mathbf{F})'(\boldsymbol{w} + (\boldsymbol{\Sigma}\boldsymbol{c} - \boldsymbol{w})) = \mathbf{g}. \tag{8}$$

On the other hand, $\boldsymbol{w} \in \mathcal{M}(\boldsymbol{\Sigma}^{-1}\mathbf{F})$, hence $\boldsymbol{w} = \boldsymbol{\Sigma}^{-1}\mathbf{F}\boldsymbol{u}$ for some $\boldsymbol{u} \in R^m$. Thus, according to (8)

$$\mathbf{g} = \mathbf{F}'\boldsymbol{\Sigma}^{-2}\mathbf{F}\boldsymbol{u} = \mathbf{M}\boldsymbol{u}. \; \square$$

PROOF OF PROPOSITION II.2. Let $\mathbf{\Sigma}$ be defined as in the previous proof.

The estimate $\mathbf{c}'\mathbf{y}$ is unbiassed exactly if

$$\mathbf{c}'\mathbf{F}\alpha = \mathrm{E}_\vartheta(\mathbf{c}'\mathbf{y}) = g(\vartheta) = \mathbf{g}'\alpha\,; \quad (\alpha \in R^m)$$

which is equivalent to the equality

$$\mathbf{g} = \mathbf{F}'\mathbf{c} \tag{9}$$

and, according to Proposition II.3, it is equivalent to $\mathbf{g} \in \mathcal{M}(\mathbf{M})$. Let $\mathbf{u}$ be a vector which is a solution of $\mathbf{g} = \mathbf{M}\mathbf{u}$, and let $\mathbf{z}_g \in \mathcal{M}(\mathbf{M})$ be obtained by projecting $\mathbf{u}$ onto $\mathcal{M}(\mathbf{M})$. Hence $(\mathbf{u} - \mathbf{z}_g)'\mathbf{M}\mathbf{v} = 0$ for every vector $\mathbf{v} \in R^m$. Particularly, taking $\mathbf{v} = \mathbf{M}(\mathbf{u} - \mathbf{z}_g)$ we obtain $\|\mathbf{M}(\mathbf{u} - \mathbf{z}_g)\|^2 = 0$, thus $\mathbf{M}\mathbf{z}_g = \mathbf{M}\mathbf{u} = \mathbf{g}$. If $\mathbf{z}$ is any vector such that $\mathbf{z} \in \mathcal{M}(\mathbf{M})$ and that $\mathbf{M}\mathbf{z} = \mathbf{g}$, then $\mathbf{z} - \mathbf{z}_g = \mathbf{M}\mathbf{t}$ for some $\mathbf{t} \in R^m$, and $\mathbf{M}(\mathbf{z} - \mathbf{z}_g) = 0$. It follows that

$$\|\mathbf{z} - \mathbf{z}_g\|^2 = [\mathbf{M}(\mathbf{z} - \mathbf{z}_g)]'\mathbf{t} = 0$$

thus $\mathbf{z} = \mathbf{z}_g$.

Since the estimate $\mathbf{c}'\mathbf{y}$ is unbiassed, we obtain for every $\alpha \in R^m$, using the Schwarz inequality

$$\begin{aligned}(\mathbf{g}'\alpha)^2 &= [\mathrm{E}_\vartheta(\mathbf{c}'\mathbf{y})]^2 = (\mathbf{c}'\mathbf{F}\alpha)^2 = (\mathbf{c}'\mathbf{\Sigma}(\mathbf{\Sigma}^{-1}\mathbf{F}\alpha))^2 \\ &\leqslant (\mathbf{c}'\mathbf{\Sigma}^2\mathbf{c})(\alpha'\mathbf{M}\alpha) = \mathrm{D}(\mathbf{c}'\mathbf{y})(\alpha'\mathbf{M}\alpha).\end{aligned}$$

Hence

$$\mathrm{D}(\mathbf{c}'\mathbf{y}) \geqslant \sup\left\{\frac{(\mathbf{g}'\alpha)^2}{\alpha'\mathbf{M}\alpha} : \alpha \in R^m,\ \mathbf{M}\alpha \neq \mathbf{0}\right\}. \tag{10}$$

Take $\mathbf{c}^* = \mathbf{\Sigma}^{-2}\mathbf{F}\mathbf{z}_g$. The estimate

$$\hat{g} \equiv \mathbf{c}^{*\prime}\mathbf{y} = \mathbf{z}_g'\mathbf{F}'\mathbf{\Sigma}^{-2}\mathbf{y} = \sum_{i=1}^{N} \mathbf{z}_g'\mathbf{f}(x_i)\sigma^{-2}(x_i)y(x_i)$$

is unbiassed, and

$$\mathrm{D}(\hat{g}) = \mathbf{z}_g'\mathbf{M}\mathbf{z}_g.$$

Moreover,

$$\frac{(\mathbf{g}'\mathbf{z}_g)^2}{\mathbf{z}_g'\mathbf{M}\mathbf{z}_g} = \mathbf{z}_g'\mathbf{M}\mathbf{z}_g.$$

After comparing this expression with Eq. (10) we obtain

$$\mathrm{D}(\hat{g}) = \sup\left\{\frac{(\mathbf{g}'\boldsymbol{\alpha})^2}{\boldsymbol{\alpha}'\mathbf{M}\boldsymbol{\alpha}} : \boldsymbol{\alpha} \in R^m,\ \mathbf{M}\boldsymbol{\alpha} \neq \mathbf{0}\right\}. \square$$

PROPOSITION II.4. *If the matrix* $\mathbf{M}$ *is nonsingular, then the BLUE for g is*

$$\hat{g} = \sum_{i=1}^{N} \mathbf{g}'\mathbf{M}^{-1}\mathbf{f}(x_i)\sigma^{-2}(x_i)y(x_i).$$

PROOF. Evidently, $\mathbf{z}_g = \mathbf{M}^{-1}\mathbf{g}$. $\square$

COROLLARY. *If* $\mathbf{M}$ *is nonsingular, then the vector of the BLUE-s for* $\alpha_1, \ldots, \alpha_m$ *is equal to*

$$\hat{\boldsymbol{\alpha}} = \mathbf{M}^{-1} \sum_{i=1}^{N} \mathbf{f}(x_i)\sigma^{-2}(x_i)y(x_i) \tag{11}$$

and its covariance matrix is

$$\mathbf{D}(\hat{\boldsymbol{\alpha}}) = \mathbf{M}^{-1}. \tag{12}$$

PROOF. Write $\alpha_i = \mathbf{e}_i'\boldsymbol{\alpha}$, the vector $\mathbf{e}_i$ being the i-th column of the identity $m \times m$ matrix. The estimate $\hat{\alpha}_i$ is obtained setting $\mathbf{e}_i$ into the expression for $\hat{g}$ in Proposition II.4.

Further

$$\begin{aligned}\{\mathbf{D}(\hat{\boldsymbol{\alpha}})\}_{ij} &= \mathrm{E}_\theta[(\hat{\alpha}_i - \alpha_i)(\hat{\alpha}_j - \alpha_j)] \\ &= \mathbf{e}_i'\mathbf{M}^{-1}\mathbf{F}'\boldsymbol{\Sigma}^{-2}\mathbf{D}(\mathbf{y})\boldsymbol{\Sigma}^{-2}\mathbf{F}\mathbf{M}^{-1}\mathbf{e}_j \\ &= \mathbf{e}_i'\mathbf{M}^{-1}\mathbf{F}'\boldsymbol{\Sigma}^{-2}\mathbf{F}\mathbf{M}^{-1}\mathbf{e}_j \\ &= \mathbf{e}_i'\mathbf{M}^{-1}\mathbf{e}_j = \{\mathbf{M}^{-1}\}_{ij}. \square\end{aligned}$$

Let $\mathbf{A}$ be a matrix. Any solution $\mathbf{A}^-$ of the equation

$$\mathbf{A}\mathbf{A}^-\mathbf{A} = \mathbf{A}$$

is said to be a g-inverse of $\mathbf{A}$. g-Inverse matrices are treated in detail in [107].

Denote

$$\mathcal{M}(\mathbf{A}) = \{\mathbf{A}\mathbf{u} : \mathbf{u} \in R^k\},$$

$$\mathcal{N}(\mathbf{A}) = \{\boldsymbol{u} : \boldsymbol{u} \in R^k, \mathbf{A}\boldsymbol{u} = \boldsymbol{0}\}, \tag{13}$$

where k represents the number of columns of $\mathbf{A}$.

PROPOSITION II.5. *Let* $\mathbf{A}$ *be a symmetric matrix. A vector* $\mathbf{u}$ *is an element of* $\mathcal{N}(\mathbf{A})$ *if and only if it is orthogonal to* $\mathcal{M}(\mathbf{A})$.

Every $\mathbf{u} \in \mathcal{M}(\mathbf{A})$ *can be associated a unique* $\mathbf{w}_u \in \mathcal{M}(\mathbf{A})$, *such that* $\mathbf{A}\mathbf{w}_u = \mathbf{u}$. *The matrix* $\mathbf{A}^+$ *defined by*

$$\begin{aligned} \mathbf{A}^+\boldsymbol{u} &= \boldsymbol{0} \quad \text{if } \boldsymbol{u} \in \mathcal{N}(\mathbf{A}) \\ &= \mathbf{w}_u \text{ if } \boldsymbol{u} \in \mathcal{M}(\mathbf{A}) \end{aligned}$$

is a g-inverse of $\mathbf{A}$. *Moreover,*

$$\mathbf{A}^+\mathbf{A}\mathbf{A}^+ = \mathbf{A}^+.$$

PROOF. Evidently, $\boldsymbol{u} \in \mathcal{N}(\mathbf{A}) \Leftrightarrow \boldsymbol{v}'\mathbf{A}\boldsymbol{u} = \boldsymbol{0}$ for every $\boldsymbol{v} \in R^m \Leftrightarrow \boldsymbol{u}$ is orthogonal to $\mathcal{M}(\mathbf{A})$. It follows that the linear operator defined by $\mathbf{A}$ maps $\mathcal{M}(\mathbf{A})$ onto $\mathcal{M}(\mathbf{A})$ and this mapping is one-to-one. For any $\boldsymbol{u} \in \mathcal{N}(\mathbf{A})$ we have

$$\mathbf{A}\mathbf{A}^+\mathbf{A}\boldsymbol{u} = \boldsymbol{0} = \mathbf{A}\boldsymbol{u}$$

and for any $\boldsymbol{u} \in \mathcal{M}(\mathbf{A})$

$$\mathbf{A}\mathbf{A}^+\mathbf{A}\boldsymbol{u} = \mathbf{A}\mathbf{w}_{\mathbf{A}u} = \mathbf{A}\boldsymbol{u}.$$

Thus $\mathbf{A}^+$ is a g-inverse of $\mathbf{A}$. Further, $\boldsymbol{u} \in \mathcal{N}(\mathbf{A})$ implies $\mathbf{A}^+\mathbf{A}\mathbf{A}^+\boldsymbol{u} = \boldsymbol{0}$, and $\boldsymbol{u} \in \mathcal{M}(\mathbf{A})$ implies $\mathbf{A}^+\mathbf{A}\mathbf{A}^+\mathbf{u} = \mathbf{A}^+\mathbf{A}\mathbf{w}_u = \mathbf{A}^+\boldsymbol{u}$. □

COROLLARY. *There is at least one g-inverse of the information matrix.*

Notice that another expression for a g-inverse matrix is given in Proposition II.9.

PROPOSITION II.6. *If* g *is an estimable functional, then*

$$\hat{g} = \mathbf{g}'\mathbf{M}^- \sum_{i=1}^{N} \mathbf{f}(x_i)\sigma^{-2}(x_i)y(x_i) \tag{14}$$

is the BLUE for g, regardless of the choice of the g-inverse $\mathbf{M}^-$. *Its variance can be expressed as*

$$\begin{aligned} D(\hat{g}) &= \boldsymbol{g}'\mathbf{M}^-\boldsymbol{g} \\ &= \sup\{2\boldsymbol{g}'\boldsymbol{\alpha} - \boldsymbol{\alpha}'\mathbf{M}\boldsymbol{\alpha} : \boldsymbol{\alpha} \in R^m\}. \end{aligned} \tag{15}$$

PROOF. According to Proposition II.2 we have $\boldsymbol{g} = \mathbf{M}\boldsymbol{u}$ for some $\boldsymbol{u} \in R^m$. Hence from Proposition II.5 it follows that

$$\mathbf{M}(\mathbf{M}^+\boldsymbol{g}) = \mathbf{M}\mathbf{M}^+\mathbf{M}\boldsymbol{u} = \mathbf{M}\boldsymbol{u} = \boldsymbol{g} \tag{16}$$

and that $\mathbf{M}^+\boldsymbol{g} \in \mathcal{M}(\mathbf{M})$. Hence $\mathbf{M}^+\boldsymbol{g}$ is equal to the vector $\boldsymbol{z}_g$ in Proposition II.2. Using Eq. (6) we obtain

$$\hat{\boldsymbol{g}} = \boldsymbol{g}'\mathbf{M}^+ \sum_{i=1}^{N} \boldsymbol{f}(x_i)\sigma^{-2}(x_i)y(x_i). \tag{17}$$

Let $\mathbf{M}^-$ be another g-inverse of $\mathbf{M}$. According to Proposition II.3, for any $\boldsymbol{v} \in R^N$ there is a vector $\boldsymbol{z} \in R^m$ such that $\mathbf{F}'\boldsymbol{v} = \mathbf{M}\boldsymbol{z}$. Hence

$$\boldsymbol{g}'\mathbf{M}^-\mathbf{F}'\boldsymbol{v} = \boldsymbol{u}'\mathbf{M}\mathbf{M}^-\mathbf{M}\boldsymbol{z} = \boldsymbol{u}'\mathbf{M}\boldsymbol{z} = \boldsymbol{u}'\mathbf{M}\mathbf{M}^+\mathbf{M}\boldsymbol{z} = \boldsymbol{g}'\mathbf{M}^+\mathbf{F}'\boldsymbol{v}.$$

Particularly, choosing $\boldsymbol{v} = (\sigma^{-2}(x_1)y(x_1), \ldots, \sigma^{-2}(x_N)y(x_N))'$ we obtain that

$$\hat{g} = \boldsymbol{g}'\mathbf{M}^- \sum_{i=1}^{N} \boldsymbol{f}(x_i)\sigma^{-2}(x_i)y(x_i). \tag{18}$$

Further, according to Eq. (7)

$$\begin{aligned} D(\hat{g}) &= \boldsymbol{z}_g'\mathbf{M}\boldsymbol{z}_g \\ &= \boldsymbol{z}_g'\mathbf{M}\mathbf{M}^-\mathbf{M}\boldsymbol{z}_g \\ &= \boldsymbol{g}'\mathbf{M}^-\boldsymbol{g}. \end{aligned}$$

From the definition of $\mathbf{M}^+$ in Proposition II.5 it follows that $\boldsymbol{u} \in \mathcal{M}(\mathbf{M})$ implies $\mathbf{M}\mathbf{M}^+\boldsymbol{u} = \boldsymbol{u}$, and $\boldsymbol{u} \in \mathcal{N}(\mathbf{M})$ implies $\mathbf{M}\mathbf{M}^+\boldsymbol{u} = \mathbf{0}$. Since $\mathcal{N}(\mathbf{M}) = \mathcal{M}^\perp(\mathbf{M})$, it follows that $\mathbf{M}\mathbf{M}^+$ is the orthogonal projector onto $\mathcal{M}(\mathbf{M})$. Hence

$$0 \leq \sum_{i=1}^{N} (\mathbf{M}^+\boldsymbol{g} - \boldsymbol{\alpha})'\boldsymbol{f}(x_i)\sigma^{-2}(x_i)\boldsymbol{f}'(x_i)(\mathbf{M}^+\boldsymbol{g} - \boldsymbol{\alpha})$$

$$= \mathbf{g}'\mathbf{M}^+\mathbf{g} - 2\mathbf{g}'\mathbf{M}^+\mathbf{M}\boldsymbol{\alpha} + \boldsymbol{\alpha}'\mathbf{M}\boldsymbol{\alpha}$$
$$= \mathbf{g}'\mathbf{M}^+\mathbf{g} - 2\mathbf{g}'\boldsymbol{\alpha} + \boldsymbol{\alpha}'\mathbf{M}\boldsymbol{\alpha}.$$

Therefore, $\mathbf{g}'\mathbf{M}^+\mathbf{g} \geq 2\mathbf{g}'\boldsymbol{\alpha} - \boldsymbol{\alpha}'\mathbf{M}\boldsymbol{\alpha}$ for every $\boldsymbol{\alpha} \in R^m$, with the equality sign if $\boldsymbol{\alpha} = \mathbf{M}^+\mathbf{g}$. Hence the last equality in Eq. (15) has been proven. □

COROLLARY. *If g_1, g_2 are estimable linear functionals, then the covariance of their BLUE-s is*

$$\operatorname{cov}(\hat{g}_1, \hat{g}_2) = \mathbf{g}_1'\mathbf{M}^-\mathbf{g}_2. \tag{19}$$

$\mathbf{M}^-$ *being an arbitrary g-inverse of* $\mathbf{M}$.

PROOF. Use that

$$\operatorname{cov}(\hat{g}_1, \hat{g}_2) = \tfrac{1}{4}[D(\hat{g}_1 + \hat{g}_2) - D(\hat{g}_1 - \hat{g}_2)]. \ \square$$

EXERCISE II.1. Suppose that $\sigma(x_i) = 1$; $(i = 1, \ldots, N)$. Prove that a linear functional g is estimable if and only if there is a unique vector $\mathbf{c}^* \in \mathcal{M}(\mathbf{F})$ such that $\mathbf{g} = \mathbf{F}'\mathbf{c}^*$. Prove that $\mathbf{c}^{*\prime}\mathbf{y}$ is the BLUE for g.

HINT. Take $\mathbf{c}^* = \mathbf{F}\mathbf{z}_g$. Use Proposition II.3 and its proof.

EXERCISE II.2. Prove the statement: If $\hat{g}_1, \ldots, \hat{g}_r$ are the BLUE-s for $g_1, \ldots, g_r$, then $\sum_{i=1}^{r} \beta_i \hat{g}_i$ is the BLUE for $\sum_{i=1}^{r} \beta_i g_i$.

HINT. Use Proposition II.2, Eq. (6) and the properties of the vector $\mathbf{z}_g$.

EXERCISE II.3. Suppose that $\det \mathbf{M} \neq 0$. Denote

$$Q(\boldsymbol{\alpha}) \equiv \sum_{i=1}^{N} \sigma^{-2}(x_i)[y(x_i) - \mathbf{f}'(x_i)\boldsymbol{\alpha}]^2$$

(the sum of squares). Prove that the unique solution of the equation

$$Q(\hat{\boldsymbol{\alpha}}) = \min_{\boldsymbol{\alpha} \in R^m} Q(\boldsymbol{\alpha})$$

is the vector $\hat{\boldsymbol{\alpha}}$ defined in Eq. (11) (the least-squares estimate).

HINT. Solve the equations $\partial Q/\partial \alpha_i = 0$; $(i = 1, \ldots, m)$ and express the matrix of second-order derivatives of $Q(\boldsymbol{\alpha})$.

Let us consider (asymptotic) designs (Definition II.3) instead of fixed-size designs.

DEFINITION II.5. *The (normed) information matrix of a design* ξ *is the* $m \times m$ *matrix*

$$\mathbf{M}(\xi) = \sum_{x \in X} \mathbf{f}(x)\mathbf{f}'(x)\sigma^{-2}(x)\xi(x). \tag{20}$$

Let ξ be the design associated with the fixed-size design $x_1, \ldots, x_N$ (Definition II.3). Then evidently

$$\mathbf{M} = N\mathbf{M}(\xi). \tag{21}$$

The following proposition is obtained by substituting $N\mathbf{M}(\xi)$ for $\mathbf{M}$ in Propositions II.1—II.5.

PROPOSITION II.7.

(*a*) *A linear functional g is estimable under the design* ξ *if and only if*

$$\mathbf{g} \in \mathcal{M}[\mathbf{M}(\xi)]. \tag{22}$$

Then there is a unique vector $\mathbf{z}_g \in \mathcal{M}[\mathbf{M}(\xi)]$ *such that* $\mathbf{g} = \mathbf{M}(\xi)\mathbf{z}_g$. *The variance of the BLUE for g under the design* ξ *is equal to*

$$\begin{aligned} N^{-1}\mathbf{z}_g'\mathbf{M}(\xi)\mathbf{z}_g &= N^{-1}\mathbf{g}'\mathbf{M}^-(\xi)\mathbf{g} \\ &= N^{-1} \sup\left\{\frac{(\mathbf{g}'\boldsymbol{\alpha})^2}{\boldsymbol{\alpha}'\mathbf{M}(\xi)\boldsymbol{\alpha}} : \boldsymbol{\alpha} \in R^m, \mathbf{M}(\xi)\boldsymbol{\alpha} \neq \mathbf{0}\right\} \\ &= N^{-1} \sup\{2\mathbf{g}'\boldsymbol{\alpha} - \boldsymbol{\alpha}'\mathbf{M}(\xi)\boldsymbol{\alpha} : \boldsymbol{\alpha} \in R^m\}. \end{aligned} \tag{23}$$

(*b*) *If* g_1, g_2 *are estimable functionals, then* $N^{-1}\mathbf{g}_1'\mathbf{M}^-(\xi)\mathbf{g}_2$ *is the covariance of the BLUE-s* $\hat{g}_1, \hat{g}_2$.

REMARK. For an arbitrary design ξ and an arbitrary linear functional g we define

$$\operatorname{var}_\xi g \equiv \sup\left\{\frac{(\mathbf{g}'\boldsymbol{\alpha})^2}{\boldsymbol{\alpha}'\mathbf{M}(\xi)\boldsymbol{\alpha}} : \boldsymbol{\alpha} \in R^m, \mathbf{M}(\xi)\boldsymbol{\alpha} \neq \mathbf{0}\right\}. \tag{24}$$

Evidently,

$$\begin{aligned} \operatorname{var}_\xi g &= \sup\{2\mathbf{g}'\boldsymbol{\alpha} - \boldsymbol{\alpha}'\mathbf{M}(\xi)\boldsymbol{\alpha} : \boldsymbol{\alpha} \in R^m\} \\ &= \mathbf{g}'\mathbf{M}^-(\xi)\mathbf{g}, \qquad \text{if } \mathbf{g} \in \mathcal{M}[\mathbf{M}(\xi)] \\ &= \infty, \qquad \text{if } \mathbf{g} \notin \mathcal{M}[\mathbf{M}(\xi)]. \end{aligned}$$

We define further when $g_1, g_2 \in \mathcal{M}[\mathbf{M}(\xi)]$

$$\text{cov}_\xi(g_1, g_2) \equiv \mathbf{g}_1'\mathbf{M}^-(\xi)\mathbf{g}_2 . \tag{25}$$

EXERCISE II.4. Let ξ be a design in the regression model (X, Θ, σ). Prove that

i) a linear functional g is estimable under ξ if and only if there is $\vartheta_g \in \Theta$ such that

$$g(\vartheta) = \sum_{x \in X} \vartheta(x)\vartheta_g(x)\sigma^{-2}(x)\xi(x); \quad (\vartheta \in \Theta),$$

ii) the BLUE is

$$\hat{g} = \sum_{x \in X} \vartheta_g(x)y(x)\sigma^{-2}(x)\xi(x)$$

and its variance is

$$\text{var}_\xi g = \sum_{x \in X} \vartheta_g^2(x)\sigma^{-2}(x)\xi(x).$$

iii) The function ϑ_g is unique if $\det \mathbf{M}(\xi) \neq 0$.

HINT. Define $\vartheta_g(x) \equiv \mathbf{z}_g'\mathbf{f}(x)$.

EXERCISE II.5. Let (X, Θ_1, σ), (X, Θ_2, σ) be two regression models. Suppose that $\Theta_1 \subset \Theta_2$, and let ξ be a design in both models. Let g be a linear functional on Θ_2 and let $\tilde{g}$ be the restriction of g to the set Θ_1. Prove that

$$\text{var}_\xi \tilde{g} \leq \text{var}_\xi g .$$

HINT. Use Proposition II.1.

EXERCISE II.6. Specify the sets X, Θ and the function σ in the following experiment (the statistical interpolation of a polynomial). The values of an m-degree polynomial having unknown coefficients can be observed independently at N points $x_1, \ldots, x_N \in \langle 0, 1 \rangle$. Suppose that the variances of the observed variables are constant. Write the information matrix $\mathbf{M}$, and for the case $N = m+1$, $\det \mathbf{M} \neq 0$, express the determinant of the covariance matrix of the BLUE-s.

HINT. Use the known expression for the determinant of Vandermond, cf. [106], Chapter 1.13.

EXERCISE II.7. Prove that the g-inverse $\mathbf{M}^+$ of an information matrix $\mathbf{M}$ is positive semidefinite.

HINT. Use Proposition II.5.

EXERCISE II.8. Consider a block experiment (the model (g) in Chapter II.2). Let ξ be a design. Denote by $\mathbf{Z}$ the "incidence matrix" defined by

$$\{\mathbf{Z}\}_{ij} \equiv \xi(i, j); \quad (i = 1, \ldots, v, \ j = 1, \ldots, b).$$

Define

$$r_i \equiv \sum_{l=1}^{b} \{\mathbf{Z}\}_{il}, \quad k_j \equiv \sum_{l=1}^{v} \{\mathbf{Z}\}_{lj},$$

$$\boldsymbol{r}' = (r_1, \ldots, r_v), \quad \boldsymbol{k}' = (k_1, \ldots, k_b),$$

$$\mathbf{R} = \operatorname{diag}(r_1, \ldots, r_v), \quad \mathbf{K} = \operatorname{diag}(k_1, \ldots, k_b).$$

Prove that

i)
$$\mathbf{M}(\xi) = \begin{pmatrix} 1, & \boldsymbol{r}', & \boldsymbol{k}' \\ \boldsymbol{r}, & \mathbf{R}, & \mathbf{Z} \\ \boldsymbol{k}, & \mathbf{Z}', & \mathbf{K} \end{pmatrix},$$

ii) If $r_i > 0$, $k_j > 0$; $(i = 1, \ldots, v, \; j = 1, \ldots, b)$,
then

$$\mathbf{M}^-(\xi) = \begin{pmatrix} 0 & \mathbf{0} \\ \mathbf{0} & \mathbf{Q}^- \end{pmatrix},$$

where

$$\mathbf{Q} \equiv \begin{pmatrix} \mathbf{R}, & \mathbf{Z} \\ \mathbf{Z}', & \mathbf{K} \end{pmatrix}$$

is a singular matrix. Prove that

$$\operatorname{cov}_\xi(\boldsymbol{h}'\boldsymbol{\alpha}, \boldsymbol{g}'\boldsymbol{\alpha}) = \boldsymbol{h}'\mathbf{C}^- \boldsymbol{g}; \quad (\boldsymbol{h}, \boldsymbol{g} \in R^v)$$

where

$$\mathbf{C} \equiv \mathbf{R} - \mathbf{Z}\mathbf{K}^{-1}\mathbf{Z}'$$

HINT. Use that $\boldsymbol{v}'\mathbf{Q} = (\boldsymbol{r}', \boldsymbol{k}')$ with $\boldsymbol{v} \equiv (1/2, \ldots, 1/2)$. Denote $\mathbf{Q}_\varepsilon \equiv \mathbf{Q} + \varepsilon\mathbf{I}$. From the equality

$$\begin{pmatrix} \mathbf{R} + \varepsilon\mathbf{I}, & \mathbf{Z} \\ \mathbf{Z}', & \mathbf{K} + \varepsilon\mathbf{I} \end{pmatrix} \begin{pmatrix} [\mathbf{Q}_\varepsilon^{-1}]_{\mathrm{I}}, & [\mathbf{Q}_\varepsilon^{-1}]_{\mathrm{II}} \\ [\mathbf{Q}_\varepsilon^{-1}]'_{\mathrm{II}}, & [\mathbf{Q}_\varepsilon^{-1}]_{\mathrm{III}} \end{pmatrix} = \begin{pmatrix} \mathbf{I}, & 0 \\ 0, & \mathbf{I} \end{pmatrix}$$

deduce that $[\mathbf{Q}^-]_{\mathrm{I}} = \lim_{\varepsilon > 0} [\mathbf{Q}_\varepsilon^{-1}]_{\mathrm{I}} = \mathbf{C}^-$. (Compare with Eqs. (43)—(45) in Chapter IV and with Proposition II.9.)

II.4. DEVIATIONS FROM THE LINEAR ESTIMATION AND FROM THE DEFINITION OF THE DESIGN

A) *Nonlinear estimates of linear functionals*

Only linear estimates have been considered so far. However, under certain conditions the BLUE-s are minimum variance unbiassed estimates also in the class of nonlinear estimates, as will be proved.

We shall write $y_i = y(x_i)$; $(i=1, ..., N)$ for conciseness of notation.

A real function

$$\tau : R^N \mapsto R$$

is said to be a *regular estimate* iff

$$\frac{\partial}{\partial \alpha_i} \int_{R^N} \tau(y_1, ..., y_N) h(y_1, ..., y_N | \alpha) \, dy_1 \dots dy_N$$
$$= \int_{R^N} \tau(y_1, ..., y_N) \frac{\partial}{\partial \alpha_i} h(y_1, ..., y_N | \alpha) \, dy_1 \dots dy_N;$$
$$(i=1, ..., m),$$

where $h(.|\alpha)$ is the normal probability density function

$$h(y_1, ..., y_N | \alpha)$$
$$= \prod_{i=1}^{N} \frac{1}{\sqrt{2\pi}\sigma_i} \exp\left\{ -\frac{1}{2} \sum_{j=1}^{N} (y_j - \mathbf{f}'(x_j)\alpha)^2 / \sigma_j^2 \right\}.$$

The regular estimate of g is unbiassed if

$$E_\vartheta(\tau) = g(\vartheta); \quad (\vartheta \in \Theta).$$

PROPOSITION II.8. (*The Rao-Cramèr inequality*)
The variance of any regular unbiassed estimate τ is bounded from below by

$$D_\vartheta(\tau) \geq \mathbf{g}'\mathbf{M}^+\mathbf{g}; \quad (\vartheta \in \Theta).$$

PROOF. Let $\vartheta = \mathbf{f}'\alpha$ be the true state. Consider the random vector

$$\mathbf{t} = \begin{pmatrix} \tau - g(\mathbf{f}'(\cdot)\alpha) \\ \sum_{j=1}^{N} (y_j - \mathbf{f}'(x_j)\alpha) f_1(x_j)/\sigma_j^2 \\ \vdots \\ \sum_{j=1}^{N} (y_j - \mathbf{f}'(x_j)\alpha) f_m(x_j)/\sigma_j^2 \end{pmatrix}.$$

Evidently $\mathrm{E}(\boldsymbol{t})=0$. Denote the covariance matrix of $\boldsymbol{t}$ by $\mathbf{D}(\boldsymbol{t})$. We can write

$$\mathbf{D}(\boldsymbol{t})=\begin{pmatrix}\mathrm{D}(\tau), & \boldsymbol{g}' \\ \boldsymbol{g}, & \mathbf{M}\end{pmatrix}.$$

This equality follows from the regularity of τ. In fact,

$$\begin{aligned}&\mathrm{E}\left[(\tau-\mathrm{E}(\tau))\left(\sum_j (y_j-\boldsymbol{f}'(x_j)\boldsymbol{\alpha})f_i(x_j)/\sigma_j^2\right)\right]\\ &=\int_{\boldsymbol{R}^N}\tau(y_1,\dots,y_N)\left[\frac{\partial}{\partial\alpha_i}\log h(y_1,\dots,y_N|\boldsymbol{\alpha})\right]\\ &\times h(y_1,\dots,y_N|\boldsymbol{\alpha})\,\mathrm{d}y_1\dots\mathrm{d}y_N\\ &=\frac{\partial}{\partial\alpha_i}\mathrm{E}(\tau)=\frac{\partial}{\partial\alpha_i}\sum_{k=1}^m\alpha_k g_k=g_i.\end{aligned}$$

Construct a $(m+1)\times(m+1)$ matrix

$$\mathbf{Q}\equiv\begin{pmatrix}1, & -\boldsymbol{g}'\mathbf{M}^+ \\ \mathbf{0}, & \mathbf{I}\end{pmatrix}$$

and consider the product

$$\mathbf{Q}\mathbf{D}(\boldsymbol{t})\mathbf{Q}'=\begin{pmatrix}\mathrm{D}(\tau)-\boldsymbol{g}'\mathbf{M}^+\boldsymbol{g}, & \boldsymbol{g}'-\boldsymbol{g}'\mathbf{M}^+\mathbf{M} \\ \boldsymbol{g}-\mathbf{M}\mathbf{M}^+\boldsymbol{g}, & \mathbf{M}\end{pmatrix}.$$

The matrix $\mathbf{Q}\mathbf{D}(\boldsymbol{t})\mathbf{Q}'$ is positive semidefinite. Thus

$$\mathrm{D}(\tau)-\boldsymbol{g}'\mathbf{M}^+\boldsymbol{g}\geq 0. \square$$

REMARK. Nonlinear estimates can be better than the BLUE-s if instead of the variance another measure of dispersion is minimalized. In [41] estimates of $\alpha_1, \dots, \alpha_m$ are considered yielding the minimum determinant of the covariance matrix of the estimates. Such estimates are nonlinear. However, the obtained minimum determinant depends on the unknown values of $\alpha_1, \dots, \alpha_m$, therefore the estimates cannot be used for the construction of a priori designs.

B) *The nonlinear estimation of nonlinear functionals*

We can estimate without bias also nonlinear functionals defined on Θ. The corresponding theory differs from the presented one and it is outlined in Chapter VII (for the infinite-dimensional model). Unbiassed estimates of nonlinear functionals have variances depending on the unknown state ϑ. However, estimating homogeneous polynomial functionals, the variance under the hypothesis $\vartheta = 0$ is decisive, and therefore can be used for comparing experimental designs (see Chapter IV.3). A study of the properties of minimum variance unbiassed estimates in the regression model is contained in [72, 73].

C) *Robust estimates*

As shown in Exercise II.3, the BLUE for $\boldsymbol{\alpha}$ is equal to the least-squares estimate. It is known that this estimate is influenced by outliers, i.e. it is not robust against deviations from normality. Therefore, other more robust estimates are used sometimes.

An *M-estimate* of $\boldsymbol{\alpha}$ is obtained if we minimize the sum

$$\sum_{i=1}^{N} \varrho[y(x_i) - \mathbf{f}'(x_i)\boldsymbol{\alpha}],$$

where ϱ is some (usually convex) function. Least-squares estimates and maximum likelihood estimates belong to the class of M-estimates. Under some regularity conditions it has been proven that the M-estimate of $\boldsymbol{\alpha}$ is asymptotically normal (when the size N tends to infinity) with the mean $\boldsymbol{\alpha}$ and the covariance matrix

$$k\mathbf{M}^{-1},$$

where $\mathbf{M}$ is the information matrix and k is a coefficient not depending on the used design (cf. [89, 90]).

An *L-estimate* of $\boldsymbol{\alpha}$ has the form

$$\boldsymbol{\alpha}^* = \sum_{i=1}^{N} \lambda_i y^{(i)},$$

where $y^{(1)} \ldots y^{(N)}$ is the ordered sequence $y(x_1), \ldots, y(x_N)$ and $\lambda_1, \ldots, \lambda_N$ are weight coefficients depending on the probability distribution of y.

Other robust estimates of $\boldsymbol{\alpha}$ are *R-estimates* which are based on rank tests (cf. [91]).

Both, L-estimates and R-estimates have similar asymptotical properties as M-estimates. That means, the asymptotical covariance matrix can be factorized, and the only factor depending on the design is the information matrix $\mathbf{M}$ (cf. [89, 92]).

As a consequence, the design theory presented in this book can be used also for asymptotic robust estimates.

D) *Ridge estimates*

PROPOSITION II.9. *If* $\mathbf{A}$ *is a symmetric matrix, then there is the limit*

$$\lim_{k \searrow 0} (\mathbf{A} + k\mathbf{I})^{-1}$$

and it is a g-inverse of $\mathbf{A}$. *Moreover,*

$$\lim_{k \searrow 0} (\mathbf{A} + k\mathbf{I})^{-1}\mathbf{A} \lim_{k \searrow 0} (\mathbf{A} + k\mathbf{I})^{-1} = \lim_{k \searrow 0} (\mathbf{A} + k\mathbf{I})^{-1}.$$

PROOF. For any $\boldsymbol{u} \in R^m$ and $k > 0$ we have

$$\|(\mathbf{A} + k\mathbf{I})^{-1}\mathbf{A}\boldsymbol{u}\| \leqslant \|\boldsymbol{u}\|.$$

Hence

$$\mathbf{A}\boldsymbol{u} = \lim_{k \searrow 0} \mathbf{A}(\mathbf{A} + k\mathbf{I})^{-1}\mathbf{A}\boldsymbol{u} + \lim_{k \searrow 0} k(\mathbf{A} + k\mathbf{I})^{-1}\mathbf{A}\boldsymbol{u}$$

$$= \mathbf{A}[\lim_{k \searrow 0} (\mathbf{A} + k\mathbf{I})^{-1}]\mathbf{A}\boldsymbol{u}.$$

Further, for every vector $\boldsymbol{u}$ we have

$$\lim_{k \searrow 0} [\mathbf{A} + k\mathbf{I}]^{-1}\boldsymbol{u}$$

$$= \lim_{k \searrow 0} [\mathbf{A} + k\mathbf{I}]^{-1} \lim_{k \searrow 0} [\mathbf{A} + k\mathbf{I}] \lim_{k \searrow 0} [\mathbf{A} + k\mathbf{I}]^{-1}\boldsymbol{u}. \square$$

It follows from Proposition II.9 and Eq. (14) that the BLUE

$$\hat{g} = \mathbf{g}'\mathbf{M}^{-}\mathbf{F}'\mathbf{\Sigma}^{-1}\boldsymbol{y}$$

can be approximated by

$$\mathbf{g}'(\mathbf{M} + k\mathbf{I})^{-1}\mathbf{F}'\mathbf{\Sigma}^{-2}\boldsymbol{y}$$

for some small k. The expression

$$\hat{\alpha}(k) \equiv (\mathbf{M} + k\mathbf{I})^{-1}\mathbf{F}'\mathbf{\Sigma}^{-2}\mathbf{y}$$

can be considered as an estimate of α also in the case when $\mathbf{M}$ is singular. Estimates of that form are called ridge estimates. They are linear but biassed. We have

$$\mathbf{D}(\hat{\alpha}(k)) = (\mathbf{M} + k\mathbf{I})^{-1}\mathbf{M}(\mathbf{M} + k\mathbf{I})^{-1} \leq \mathbf{M}^{-1} = \mathbf{D}(\hat{\alpha})$$

hence ridge estimates have smaller variances than the BLUE-s. We shall return to the ridge estimates later, when considering singular designs.

E) *Note*

A decision-theoretic approach could be adapted to the problem of estimation and design as in [11], that is one could introduce a loss function and the corresponding risk function and use a prior distribution of the unknown parameters $\alpha_1, \ldots, \alpha_m$ (cf. e.g. [94]). However, this approach will not be used here, with a certain exception in Propositions II.13 and II.14, because of the subjectivity of the choice of the loss function and of the prior distribution.

F) *Designs with fixed cost*

The total cost (total expense) for observations is fixed in some cases instead of the size N. Let C be this total cost, and let C_x be the cost per one observation at the point $x \in X$. The information matrix $\mathbf{M}$ of a design $x_1, \ldots, x_N$ can be expressed as

$$\mathbf{M} = C\tilde{\mathbf{M}}(\xi),$$

where $\xi(x) \equiv N_x C_x / C$, $(x \in X)$ is the relative cost of observations in the trial x, and the "normed" information matrix is

$$\tilde{\mathbf{M}}(\xi) \equiv \sum_{x \in X} \mathbf{f}(x)\mathbf{f}'(x)\tilde{\sigma}^{-2}(x)\xi(x)$$

with $\tilde{\sigma}^2(x) \equiv \sigma^2(x)C_x$. The matrix $\tilde{\mathbf{M}}(\xi)$ is similar to the matrix $\mathbf{M}(\xi)$ in Eq. (20). Hence the design theory remains unchanged.

II.5. DEVIATIONS FROM THE MODEL OF THE EXPERIMENT

This chapter contains some remarks on possible deviations from the linear regression model with uncorrelated observations.

II.5.1. *The Nonlinear Regression Model*

In the regression model the set of states is a linear space, which means that any $\vartheta \in \Theta$ can be parametrized linearly. However, in some experiments there is no unconstrained linear parametrization of the function $x \in X \mapsto \mathrm{E}[y(x)]$, or, which is equivalent, the set Θ is no more a linear space.

Observe, however, that if the estimated functional $g: \Theta \mapsto R$ is still linear (i.e. if for any $\vartheta_1, \vartheta_2 \in \Theta$, which are such that $\beta_1\vartheta_1 + \beta_2\vartheta_2 \in \Theta$, the equality

$$g(\beta_1\vartheta_1 + \beta_2\vartheta_2) = \beta_1 g(\vartheta_1) + \beta_2 g(\vartheta_2)$$

is valid), then the situation is essentially the same as in the linear model. Indeed, it is sufficient to extend g linearly on the linear space $\mathscr{L}(\Theta)$ spanned by Θ, and to estimate this extended functional, say $\tilde{g}$, linearly in the linear model $(X, \mathscr{L}(\Theta), \sigma)$, since a linear estimate is unbiassed for g if and only if it is unbiassed for $\tilde{g}$.

However, in nonlinear models important functionals are usually nonlinear. Usually, under a nonlinear regression model (with uncorrelated observations) we understand a regression model (X, Θ, σ) with

$$\Theta = \{\eta(\cdot, \alpha): \alpha \in U\}.$$

Here U is an open subset of R^m, and the function

$$\eta: (x, \alpha) \in X \times U \mapsto \eta(x, \alpha)$$

is continuous in x, and it is nonlinear and twice continuously differentiable in the parameters $\alpha_1, \ldots, \alpha_m$ (see the Example 4 in Chapter I).

In some cases the model can be linearized locally. For this, some preliminary estimates $\alpha_1^{(0)}, \ldots, \alpha_m^{(0)}$ must be obtained, and the function $\alpha \in R^m \mapsto \eta(x, \alpha)$ is approximated by the linear part of the Taylor expansion at $\alpha^{(0)}$. In the obtained *linearized model* the differences

$(\Delta\alpha)_i \equiv \alpha_i - \alpha_i^{(0)}$; $(i=1, \ldots, m)$ are estimated. Because of the linearization, the estimates $(\hat{\Delta\alpha})_1, \ldots, (\hat{\Delta\alpha})_m$ are biassed. The bias is evaluated in the following proposition.

Suppose that $\sigma^2(x)=1$; $(x \in X)$. Denote

$$\tilde{f}_i(x) \equiv \left.\frac{\partial \eta(x, \boldsymbol{\alpha})}{\partial \alpha_i}\right|_{\boldsymbol{\alpha}^{(0)}}, \tag{26}$$

$$\Psi_x(\boldsymbol{\alpha}^{(0)}, \boldsymbol{\alpha}^{(s)}, \boldsymbol{\alpha})$$
$$\equiv \frac{1}{2} \sum_{i,j=1}^{m} (\alpha_i^{(s)} - \alpha_i^{(0)}) \frac{\partial \eta^2(x, \boldsymbol{\alpha})}{\partial \alpha_i \partial \alpha_j} (\alpha_j^{(s)} - \alpha_j^{(0)}),$$

$$\mathbf{M}(\xi) = \sum_{x \in X} \tilde{\mathbf{f}}(x) \tilde{\mathbf{f}}'(x) \xi(x) \tag{27}$$

and denote the arithmetic mean of results of the observations in the trial x by $\bar{y}(x)$.

Proposition II.10 [39]. *Let $\alpha_1^{(s)}, \ldots, \alpha_m^{(s)}$ be the true values of the parameters. Suppose that $\mathbf{M}(\xi)$ is nonsingular.*

Then there is the BLUE for the vector $\boldsymbol{\Delta\alpha} = ((\Delta\alpha)_1, \ldots, (\Delta\alpha)_m)$ in the linearized model, which is equal to

$$\hat{\boldsymbol{\Delta\alpha}} = \mathbf{M}^{-1}(\xi) \sum_{x \in X} \tilde{\mathbf{f}}(x)[\bar{y}(x) - \eta(x, \boldsymbol{\alpha}^{(0)})]\xi(x). \tag{28}$$

The estimate is biassed in the nonlinearized model. There are numbers α_i^* between $\alpha_i^{(0)}$ and $\alpha_i^{(s)}$; $(i=1, \ldots, m)$ such that the bias can be expressed as

$$\mathrm{E}(\hat{\boldsymbol{\Delta\alpha}}) - \boldsymbol{\Delta\alpha} = \mathbf{M}^{-1}(\xi) \sum_{x \in X} \tilde{\mathbf{f}}(x) \Psi_x(\boldsymbol{\alpha}^{(0)}, \boldsymbol{\alpha}^{(s)}, \boldsymbol{\alpha}^*)\xi(x). \tag{29}$$

Proof. Using the Taylor formula we obtain

$$\mathrm{E}[\bar{y}(x) - \eta(x, \boldsymbol{\alpha}^{(0)})]$$
$$= \sum_{i=1}^{m} \tilde{f}_i(x)(\alpha_i^{(s)} - \alpha_i^{(0)}) + \Psi_x(\boldsymbol{\alpha}^{(0)}, \boldsymbol{\alpha}^{(s)}, \boldsymbol{\alpha}^*), \tag{30}$$

where the last term is the residual term in the Taylor formula. This term is neglected in the linearized model, and it is supposed that

$$\mathrm{E}[\bar{y}(x)-\eta(x, \boldsymbol{\alpha}^{(0)})]=\sum_{i=1}^{m} \bar{f}_i(x)(\Delta\alpha)_i. \tag{31}$$

That means that the variables $\bar{y}(x)-\eta(x, \boldsymbol{\alpha}^{(0)})$ are supposed to be observed in the linearized model, and that $(\Delta\alpha)_1, \ldots, (\Delta\alpha)_m$ are the parameters in this model. The estimates can be obtained from the expression in Eq. (11), substituting $\bar{y}(x)-\eta(x, \boldsymbol{\alpha}^{(0)})$ for $y(x)$, and taking $\sigma(x)=1$. From Eq. (30) we obtain

$$\begin{aligned}\mathrm{E}(\hat{\Delta}\boldsymbol{\alpha})=\mathbf{M}^{-1}(\xi) \sum_{x\in X} \mathbf{f}(x)[\mathbf{f}'(x)(\boldsymbol{\alpha}^{(s)}-\boldsymbol{\alpha}^{(0)}) \\ +\Psi_x(\boldsymbol{\alpha}^{(0)}, \boldsymbol{\alpha}^{(s)}, \boldsymbol{\alpha}^*)]\xi(x)=(\boldsymbol{\alpha}^{(s)}-\boldsymbol{\alpha}^{(0)}) \\ +\mathbf{M}^{-1}(\xi) \sum_{x\in X} \mathbf{f}(x)\Psi_x(\boldsymbol{\alpha}^{(0)}, \boldsymbol{\alpha}^{(s)}, \boldsymbol{\alpha}^*)\xi(x).\end{aligned}$$

The last term is the bias. □

Similarities between the right-hand sides of Eqs. (28) and (30) simplify the computation of the bias on computers.

In some cases the described linearization of the model cannot be used, because the obtained approximation would be too rough. In such a case the *least-squares estimates*

$$\hat{\boldsymbol{\alpha}}=\operatorname{Arg}\min_{\alpha\in U} \sum_{i=1}^{N} [y(x_i)-\eta(\boldsymbol{\alpha}, x_i)]^2$$

should be computed. The variances of such estimates depend strongly on the unknown value of $\boldsymbol{\alpha}$ and can no more be computed from the information matrix. For an expression of the probability density of the least-squares estimates in such a case cf. [74]. Difficulties arising in designing nonlinear regression experiments are discussed in [12].

II.5.2. *The Regression Model with Correlated Observations*

In some experiments the observed random variables $y(x_1), \ldots, y(x_N)$ are correlated. Let $\mathbf{K}$ be the covariance matrix of these random variables. Suppose, just as before, that

$$\mathrm{E}_\alpha[y(x_i)] = \boldsymbol{f}'(x_i)\boldsymbol{\alpha}; \quad (i=1, \ldots, N)$$

and that the parameter function $\boldsymbol{g}'\boldsymbol{\alpha}$ should be estimated. Denote by $\mathbf{M}_K$ the matrix

$$\mathbf{M}_K = \sum_{i,j=1}^{N} \boldsymbol{f}(x_i)\{\mathbf{K}^{-1}\}_{ij}\boldsymbol{f}'(x_j).$$

PROPOSITION II.11. *The parameter function $\boldsymbol{g}'\boldsymbol{\alpha}$ is estimable without bias exactly if $\boldsymbol{g} \in \mathcal{M}(\mathbf{M}_K)$. The BLUE for $\boldsymbol{g}'\boldsymbol{\alpha}$ is*

$$\boldsymbol{g}'\mathbf{M}_K^- \sum_{i,j=1}^{N} \boldsymbol{f}(x_i)\{\mathbf{K}^{-1}\}_{ij}y(x_j)$$

and its variance is

$$\boldsymbol{g}'\mathbf{M}_K^-\boldsymbol{g}.$$

The proof is a modification of the proofs of Propositions II.2 and II.6.

From Proposition II.11 it follows that the methods of estimation do not differ essentially from the methods used when the observations are uncorrelated. On the other hand, the theory and the methods of experimental design differ greatly in the correlated and uncorrelated cases. Moreover, in the case of correlated observations the design theory depends very much on the structure of the covariance matrix of the observed variables.

The robustness of designs against small perturbations in the covariance matrix from the uncorrelated case is considered in [78, 79]. Different approaches to the optimum design theory in the correlated case are presented, e.g. in [75—78]. We shall not consider the designs for correlated observations in this book.

II.5.3. *The Regression Model with Grouped Observations*

In some experiments not one variable but a group of variables $y^{(1)}(x)$, ..., $y^{r(x)}(x)$ is observed in one trial $x \in X$. Denote $y(x) \equiv (y^{(1)}(x), \ldots, y^{r(x)}(x))$. Further denote the covariance matrix of the random vector

$\mathbf{y}(x)$ by $\mathbf{K}(x)$. For the sake of simplicity let us suppose that there exists the inverse matrix $\mathbf{K}^{-1}(x)$ for every $x \in X$. It is supposed, just as before, that the random variables observed in different or repeated trials are uncorrelated. The mean of the observed vector is

$$E_\alpha[\mathbf{y}(x)] = \mathbf{G}'(x)\boldsymbol{\alpha}; \quad (x \in X, \boldsymbol{\alpha} \in R^m),$$

where $\mathbf{G}(x)$ is a given $m \times r(x)$ matrix and $\boldsymbol{\alpha} = (\alpha_1, \ldots, \alpha_m)$ is the vector of the unknown parameters.

Denote the design by ξ (see Definition II.3). The information matrix of the design ξ is

$$\mathbf{M}_G(\xi) = \sum_{x \in X} \mathbf{G}(x)\mathbf{K}^{-1}(x)\mathbf{G}'(x)\xi(x).$$

Given a linear function of the parameters

$$\boldsymbol{\alpha} \in R^m \mapsto \mathbf{g}'\boldsymbol{\alpha} \in R \tag{32}$$

we look for an estimate like

$$\sum_{x \in X} c(x)\bar{\mathbf{y}}(x)\xi(x),$$

where $\bar{\mathbf{y}}(x)$ is the arithmetic mean of the observed results in the trial x and $c(.)$ is an adequate function.

Proposition II.12. *The function (32) is estimable without bias under the design ξ if and only if $\mathbf{g} \in \mathcal{M}[\mathbf{M}_G(\xi)]$. The BLUE for $\mathbf{g}'\boldsymbol{\alpha}$ is equal to*

$$\mathbf{g}'\mathbf{M}_G^-(\xi) \sum_{x \in X} \mathbf{G}(x)\mathbf{K}^{-1}(x)\bar{\mathbf{y}}(x)\xi(x)$$

and its variance is proportional to

$$\mathbf{g}'\mathbf{M}_G^-(\xi)\mathbf{g}.$$

The proof is similar to the proofs of Propositions II.2 and II.6.

II.5.4. *Vaguely Defined Regression Models*

In some experiments we do not know exactly the regression model. For example, we do not know the boundary of the set X, or the possible state functions $\vartheta \in \Theta$ are not known exactly in the neighbourhood of the boundary of the set X. Often, we are not sure about the dimension of the linear space Θ (for example, the degree of the polynomial in the polynomial regression (see Example II.6) is uncertain). Therefore, besides exact data on the model there are also uncertain, vague data which even though unreliable, still have to be taken into account.

There are at least two possibilities to deal with this vagueness. One of them is to consider testing alternative models together with the parameter estimation. This will be considered in this book when two alternatives of the dimension of Θ are possible (see Chapter V.6.4 for the corresponding design). The other possibility is to express the vague knowledge using a more or less subjective, a priori probability distribution on the space of states Θ. This will be done in Proposition II.13. In correspondence with the linear structure of the regression model, it will be supposed that the a priori probability distribution is Gaussian, i.e. that a state ϑ is a sample from a Gaussian random process defined on X (about Gaussian random processes cf. [87]).

The model of the experiment used in Proposition II.13 is the following (cf. [40]). Let $\mathbf{K}^{(0)}$ be a symmetric positive, definite $m \times m$ matrix, let ϱ be a nonnegative decreasing function defined on $\langle 0, \infty)$ such that $\varrho(0)=1$. Let X be a compact subspace of R^k and let h be a nonnegative function defined on X. Let $\alpha^{(0)} \in R^m$ be a vector. Consider the m-dimensional Gaussian random process

$$\{\alpha(x) : x \in X\}$$

having the mean

$$x \in X \mapsto \alpha^{(0)} \tag{33}$$

and the covariance function

$$\begin{aligned}(x, x^*) \in X \times X \mapsto & \mathrm{E}[(\alpha(x) - \alpha^{(0)})(\alpha(x^*) - \alpha^{(0)})'] \\ & \equiv h(x)h(x^*)\varrho(\|x - x^*\|)\mathbf{K}^{(0)}.\end{aligned} \tag{34}$$

Similarly as in the regression model we denote by $f_1, \ldots, f_m$ linearly independent functions defined on X. The linear regression model is supposed to be contained in the considered model when conditioning upon $\alpha(.)$, that means it is supposed that

$$\mathrm{E}[y(x)|\boldsymbol{\alpha}(.)]=\boldsymbol{f}'(x)\boldsymbol{\alpha}(x); \quad (x\in X),$$

that

$$\begin{aligned}\operatorname{cov}[y(x), y(x^*)|\boldsymbol{\alpha}(.)]&=1, \text{ if } x=x^*\\ &=0, \text{ if } x\neq x^*\end{aligned}$$

and that $y(x)$ is distributed normally, under the condition that $\boldsymbol{\alpha}(.)$ is given.

Consider a fixed-size design $x_1, \ldots, x_N$. Denote

$$\boldsymbol{y}'=(y(x_1), \ldots, y(x_N)), \quad \mathbf{F}'=(\boldsymbol{f}(x_1), \ldots, \boldsymbol{f}(x_N)),$$

$$\mathbf{S}(x)=\begin{pmatrix} h(x)h(x_1)\varrho(\|x-x_1\|)\boldsymbol{f}'(x_1)\mathbf{K}^{(0)} \\ \cdots\cdots\cdots\cdots\cdots\cdots\cdots\cdots \\ h(x)h(x_N)\varrho(\|x-x_N\|)\boldsymbol{f}'(x_N)\mathbf{K}^{(0)}\end{pmatrix},$$

$$\{\mathbf{C}\}_{ij}=h(x_i)h(x_j)\varrho(\|x_i-x_j\|)\boldsymbol{f}'(x_i)\mathbf{K}^{(0)}\boldsymbol{f}(x_j);$$
$$(i, j=1, \ldots, N)$$

$$\mathbf{Q}(x)=\mathbf{I}-\mathbf{F}'(\mathbf{I}+\mathbf{C})^{-1}\mathbf{S}(x).$$

PROPOSITION II.13. *The random process*

$$\{\boldsymbol{\alpha}(x): x\in X\}$$

is Gaussian also a posteriori (i.e. after observing $y(x_1), \ldots, y(x_N)$) with the a posteriori mean

$$\begin{aligned} x\in X\mapsto\bar{\boldsymbol{\alpha}}(x)&\equiv\mathrm{E}\{\boldsymbol{\alpha}(x)|y, \boldsymbol{\alpha}^{(0)}\}\\ &=\mathbf{S}'(x)(\mathbf{I}+\mathbf{C})^{-1}\boldsymbol{y}+\mathbf{Q}'(x)\boldsymbol{\alpha}^{(0)} \end{aligned} \tag{35}$$

and the a posteriori covariance function

$$\begin{aligned} &(x, x^*)\in X\times X\mapsto\mathrm{E}\{(\boldsymbol{\alpha}(x)-\bar{\boldsymbol{\alpha}}(x))\\ &\times(\boldsymbol{\alpha}(x^*)-\bar{\boldsymbol{\alpha}}(x^*))|\boldsymbol{y}, \boldsymbol{\alpha}^{(0)}\}\\ &=h(x)h(x^*)\varrho(\|x-x^*\|)\mathbf{K}^{(0)}\\ &-\mathbf{S}'(x)(\mathbf{I}+\mathbf{C})^{-1}\mathbf{S}(x^*). \end{aligned} \tag{36}$$

PROOF. To prove the proposition it is necessary to find the distribution of the random vector

$$\boldsymbol{v}\equiv(\boldsymbol{y}', \boldsymbol{\alpha}'(x), \boldsymbol{\alpha}'(x^*))'$$

for fixed x, x^*. The random vector $(\boldsymbol{\alpha}'(x), \boldsymbol{\alpha}'(x^*))$ is Gaussian and $\boldsymbol{y}$ is conditionally Gaussian when $\boldsymbol{\alpha}(.)$ is given. Therefore, $\boldsymbol{v}$ is a Gaussian random vector ([88], Chapter 8). It can be verified without difficulty that

$$\mathrm{E}[y(x_i)] = \mathrm{E}\{\mathrm{E}[y(x_i)|\boldsymbol{\alpha}(.)]\} = \boldsymbol{f}'(x_i)\boldsymbol{\alpha}^{(0)}, \tag{37}$$

$$\begin{aligned}
&\operatorname{cov}[y(x_i), y(x_j)] \\
&= \mathrm{E}\{\mathrm{E}[(y(x_i) - \boldsymbol{f}'(x_i)\boldsymbol{\alpha}(x_i) + \Delta_i)(y(x_j) \\
&\quad - \boldsymbol{f}'(x_j)\boldsymbol{\alpha}(x_j) + \Delta_j)|\boldsymbol{\alpha}(.)]\} = \{\mathbf{I}\}_{ij} + \{\mathbf{C}\}_{ij},
\end{aligned}$$

where $\Delta_i \equiv \boldsymbol{f}'(x_i)\boldsymbol{\alpha}(x_i) - \boldsymbol{f}'(x_i)\boldsymbol{\alpha}^{(0)}$.

Similarly,

$$\begin{aligned}
&\operatorname{cov}[y(x_i), \boldsymbol{\alpha}(x)] \\
&= \mathrm{E}\{\mathrm{E}[(y(x_i) - \boldsymbol{f}'(x_i)\boldsymbol{\alpha}^{(0)})(\boldsymbol{\alpha}(x) - \boldsymbol{\alpha}^{(0)})|\boldsymbol{\alpha}(.)]\} \\
&= h(x)h(x_i)\varrho(\|x - x_i\|)\boldsymbol{f}'(x_i)\mathbf{K}^{(0)}, \\
&\operatorname{cov}[\boldsymbol{\alpha}(x), \boldsymbol{\alpha}(x^*)] = h(x)h(x^*)\varrho(\|x - x^*\|)\mathbf{K}^{(0)}.
\end{aligned}$$

Hence the covariance matrix of $\boldsymbol{v}$ is

$$\mathbf{D}(\boldsymbol{v}) = \begin{pmatrix} \mathbf{I}+\mathbf{C}, & \mathbf{S}(x), & \mathbf{S}(x^*) \\ \mathbf{S}'(x), & \mathbf{K}^{(0)}, & h(x)h(x^*)\varrho(\|x-x^*\|)\mathbf{K}^{(0)} \\ \mathbf{S}'(x), & h(x)h(x^*)\varrho(\|x-x^*\|)\mathbf{K}^{(0)}, & \mathbf{K}^{(0)} \end{pmatrix}. \tag{38}$$

As well known (cf. [88], Chapter 8, Eq. (8a.2.11)), if $\boldsymbol{t} = (\boldsymbol{t}_1', \boldsymbol{t}_2')$ is a Gaussian random vector with the mean $(\boldsymbol{\mu}_1', \boldsymbol{\mu}_2')$ and the covariance matrix

$$\mathbf{K} = \begin{pmatrix} \mathbf{K}_{\mathrm{I}}, & \mathbf{K}_{\mathrm{II}} \\ \mathbf{K}_{\mathrm{II}}', & \mathbf{K}_{\mathrm{III}} \end{pmatrix},$$

then the conditional distribution of the random vector $\boldsymbol{t}_2$, conditioned by $\boldsymbol{t}_1$, is Gaussian with the mean

$$\boldsymbol{\mu}_2 + \mathbf{K}_{\mathrm{II}}'\mathbf{K}_{\mathrm{I}}^-(\boldsymbol{t}_1 - \boldsymbol{\mu}_1) \tag{39}$$

and the covariance matrix

$$\mathbf{K}_{\mathrm{III}} - \mathbf{K}'_{\mathrm{II}}\mathbf{K}_{\mathrm{I}}^{-}\mathbf{K}_{\mathrm{II}}.$$

Apply these results to $\boldsymbol{t} = \boldsymbol{v}$, $\boldsymbol{t}_1 = \boldsymbol{y}$, $\boldsymbol{t}_2 = (\alpha'(x), \alpha'(x^*))'$. The expressions in Eqs. (35) and (36) will be obtained. □

The function $x \in X \mapsto \boldsymbol{f}'(x)\bar{\alpha}(x)$, where $\bar{\alpha}(x)$ is given by Eq. (35), is the "most probable" state a posteriori (i.e. after observing $y(x_1), \ldots, y(x_N)$). It also depends on the a priori most probable mean $\boldsymbol{\alpha}^{(0)}$ (Eq. (33)). We can go further and eliminate the dependence on $\boldsymbol{\alpha}^{(0)}$, supposing that also $\boldsymbol{\alpha}^{(0)}$ is a Gaussian random vector with a covariance matrix $\beta\mathbf{K}^*$ and taking $\beta \to \infty$.

This can be formulated more exactly, as done in the following Proposition II.14 which has been proven in [40].

Denote the a posteriori mean of the process $\{\alpha(x): x \in X\}$ by $\mathrm{E}\{\alpha(x)|\alpha^*, \beta, \mathbf{K}^*, \boldsymbol{y}\}$ under the assumption that $\boldsymbol{\alpha}^{(0)}$ is normally distributed $N(\alpha^*, \beta\mathbf{K}^*)$.

PROPOSITION II.14. *For every* $x \in X$

$$\lim_{\beta \to \infty} \mathrm{E}\{\alpha(x)|\alpha^*, \beta, \mathbf{K}^*, \boldsymbol{y}\} = \hat{\alpha} + \mathbf{S}'(x)(\mathbf{I} + \mathbf{C})^{-1}(\boldsymbol{y} - \mathbf{F}\hat{\alpha}), \tag{40}$$

where

$$\hat{\alpha} = [\mathbf{F}'(\mathbf{I} + \mathbf{C})^{-1}\mathbf{F}]^{-1}\mathbf{F}'(\mathbf{I} + \mathbf{C})^{-1}\boldsymbol{y}. \tag{41}$$

If $h(x)$ tends to zero for every $x \in X$, then, as seen from Eq. (41), $\hat{\alpha}$ is the BLUE in the original linear regression model with uncorrelated observations. It is because the covariance of the random process tends also to zero, as seen from Eq. (34). In this case the conditional mean given by Eq. (40) tends to $\hat{\alpha}$, hence there is no uncertainty in the model.

In the general case $\hat{\alpha}$ remains to be the BLUE, but the linear model with correlated observations with the covariance matrix $(\mathbf{I} + \mathbf{C})$ has to be considered. The matrix $\mathbf{C}$ is positive definite, hence, due to uncertainty there is a loss of information about α. The second term in Eq. (40) depends on the residual vector $\boldsymbol{y} - \mathbf{F}\hat{\alpha}$. It is large when this vector is large, i.e. when the validity of the original regression model is suspect. In general, we can say that by a choice of large values of $\{\mathbf{K}^{(0)}\}_{ii}$ we express the suspicion that the parameter α_i is superfluous in the regression model, by a large value of $h(x)$ we express the suspicion that observations in x are either impossible or of a small information content, by an outstanding decrease of the function $\varrho(t)$ we express that the state can be described by a function which differs essentially from the state of the linear regression model.

Chapter III

The Ordering of Designs and the Properties of Variances of Estimates

III.1. INTRODUCTION

As explained in Chapter II, a design in a regression model (X, Θ, σ) is an arbitrary measure ξ on X which is normed to 1 and supported by a finite set. It is useful to study the properties of *the set of all designs* in the model; this set will be denote by Ξ. Equivalently, we can study the properties of *the set of all information matrices*, $\mathfrak{M} \equiv \{\mathbf{M}(\xi) : \xi \in \Xi\}$. The set $\mathfrak{M}$ is evidently convex, and as shown, it is a compact subset of an Euclidean space. This allows, basing on the Caratheodory theorem (Proposition III.8), to restrict the number of points needed in the support of a design ξ without changing the information matrix $\mathbf{M}(\xi)$ (see Proposition II.11).

The aim of every optimalization method is to find an adequate design, i.e. an element of the rather complicated set Ξ. Fortunately, the quality of a design depends only on its information matrix, as follows from Proposition II.7. The set $\mathfrak{M}$ has a simpler structure than the set Ξ. However, the variance of an estimate is a relatively simple linear form of the g-inverse and not of the information matrix itself (see Eq. (23) in Proposition II.7),

$$\mathrm{var}_\xi g = \boldsymbol{g}'\mathbf{M}^-(\xi)\boldsymbol{g}; \quad (\boldsymbol{g} \in \mathcal{M}[\mathbf{M}(\xi)]).$$

The behaviour of $\mathrm{var}_\xi g$ is rather subtile if $\mathbf{M}(\xi)$ is singular, or nearly singular. Therefore, a detailed investigation of convexity and continuity of the mapping $\mathbf{M}(\xi) \in \mathfrak{M} \mapsto \mathrm{var}_\xi g$, is given in Chapter III.3. Two results which can help to find the design having a minimum value of $\mathrm{var}_\xi g$ for a given g, are presented in Propositions III.16 and III.17.

The natural requirement that a design should be considered universally optimal, when minimizing the value of $\mathrm{var}_\xi g$ for every linear functional g, leads to *the uniform ordering* of designs discussed in Chapter III.2, eventually to its refinement considered in Propositions III.4 and III.5.

AN IMPORTANT NOTE. In this chapter and throughout the remaining part of this book it will be supposed that

$$\sigma(x)=1; \quad (x \in X)$$

in any considered regression experiment. It is no serious restriction on generality since the results which are valid for $\sigma(x)=1$ are also valid in the general case (unless the contrary is stated). It is just necessary to set $\sigma^{-1}(x)\vartheta(x)$ and $\sigma^{-1}(x)\boldsymbol{f}(x)$ instead of the function $\vartheta(x)$ and of the vector $\boldsymbol{f}(x)$ whenever they appear. For example, if

$$\sum_{x \in X} \boldsymbol{f}(x)\boldsymbol{f}'(x)\xi(x)$$

is the information matrix in the experiment $(X, \Theta, 1)$, then by the substitution $\boldsymbol{f}(x) \mapsto \boldsymbol{f}(x)\sigma^{-1}(x)$ we obtain

$$\sum_{x \in X} \boldsymbol{f}(x)\boldsymbol{f}'(x)\sigma^{-2}(x)\xi(x),$$

which is the information matrix of ξ in the experiment (X, Θ, σ).

III.2. THE UNIFORM ORDERING OF THE SET OF DESIGNS

DEFINITION III.1. *Two designs ξ and η are equivalent (write $\xi \sim \eta$) iff for every linear functional g defined on Θ we have*

$$\mathrm{var}_\xi g = \mathrm{var}_\eta g.$$

The design ξ is uniformly not worse than η (write $\xi \lesssim \eta$) iff for every linear functional g on Θ we have

$$\mathrm{var}_\xi g \leqslant \mathrm{var}_\eta g.$$

The design ξ is uniformly better than η (write $\xi < \eta$) iff $\xi \lesssim \eta$ but ξ is not equivalent to η.

According to Eq. (24) in Chapter II, the values $\mathrm{var}_\xi g = \infty$ or

$\text{var}_\eta g = \infty$ are accepted in Definition III.1 when the functional g is not estimable under the design ξ or under the design η.

The ordering of the set Ξ according to the relation $\lesssim$ is called *the uniform ordering*. Evidently, it is a partial ordering of Ξ.

NOTES. We recall that a nonzero vector $\boldsymbol{u} \in R^m$ is an *eigenvector* of a matrix $\mathbf{A}$ if there is a number γ such that $\mathbf{A}\boldsymbol{u} = \gamma \boldsymbol{u}$. The number γ is the *eigenvalue* corresponding to the eigenvector $\boldsymbol{u}$. A positive semidefinite (respectively definite) matrix $\mathbf{A}$ has only nonnegative (respectively positive) eigenvalues, since $\gamma = \boldsymbol{u}'\mathbf{A}\boldsymbol{u}/\boldsymbol{u}'\boldsymbol{u} \geq 0$. (For the definition of positive definiteness see the note in Chapter II.3.)

Any symmetric $m \times m$ matrix has exactly m orthogonal eigenvectors $\boldsymbol{u}_1, \ldots, \boldsymbol{u}_m$ which are normed to 1 (i.e. $\boldsymbol{u}_i\boldsymbol{u}_j = 0$ if $i \neq j$, $\boldsymbol{u}_i'\boldsymbol{u}_i = 1$; cf. [106], Chapter II).

According to the notation introduced in Chapter II.3, by

$$\mathbf{A} \geq \mathbf{B}$$

we denote that the matrix $\mathbf{A} - \mathbf{B}$ is positive semidefinite and by

$$\mathbf{A} > \mathbf{B}$$

that $\mathbf{A} - \mathbf{B}$ is positive definite. (The ordering $\geq$ of the set of symmetric $m \times m$ matrices is sometimes referred to as the Loewner ordering.)

Let us return to the ordering of designs. If a linear basis is given in the space Θ (as in Chapter II), then the uniform ordering of designs can be described advantageously by the Loewner ordering of information matrices.

PROPOSITION III.1. *Two designs are equivalent,* $\xi \sim \eta$, *if and only if* $\mathbf{M}(\xi) = \mathbf{M}(\eta)$.

The proof follows from Propositions II.5 and II.8.

We remind of the notation

$$\mathcal{M}[\mathbf{M}(\xi)] \equiv \{\mathbf{M}(\xi)\boldsymbol{u}: \boldsymbol{u} \in R^m\},$$

$$\mathcal{N}[\mathbf{M}(\xi)] \equiv \{\boldsymbol{u}: \boldsymbol{u} \in R^m, \mathbf{M}(\xi)\boldsymbol{u} = \boldsymbol{0}\}.$$

PROPOSITION III.2. *If* $\mathbf{M}(\xi) \geq \mathbf{M}(\eta)$, *then* $\xi \lesssim \eta$. *Conversely, from* $\xi \lesssim \eta$ *it follows that* $\mathcal{M}[\mathbf{M}(\eta)] \subset \mathcal{M}[\mathbf{M}(\xi)]$ *and that*

$$\boldsymbol{u}'\mathbf{M}(\xi)\boldsymbol{u} \geq \boldsymbol{u}'\mathbf{M}(\eta)\boldsymbol{u}; \quad (\boldsymbol{u} \in \mathcal{M}[\mathbf{M}(\eta)]).$$

PROOF. Suppose that $\mathbf{M}(\xi) \geq \mathbf{M}(\eta)$. Denote by $\boldsymbol{u}_1, \ldots, \boldsymbol{u}_m$ the orthogonal eigenvectors of $\mathbf{M}(\eta)$, and denote by $\boldsymbol{U}$ the matrix $\boldsymbol{U} \equiv (\boldsymbol{u}_1, \ldots, \boldsymbol{u}_m)$. The matrix $\mathbf{D} \equiv \mathbf{U}'\mathbf{M}(\eta)\mathbf{U}$ is evidently diagonal, $\mathbf{D} = \text{diag}\,(d_1, \ldots, d_m)$, with nonnegative diagonal elements $d_1, \ldots, d_m$. Denote by $\mathbf{L}$ the matrix

$$\mathbf{L} = \text{diag}\,(\sqrt{d_1}, \ldots, \sqrt{d_m})\mathbf{U}^{-1}.$$

Evidently, $\mathbf{M}(\eta) = \mathbf{U}'^{-1}\mathbf{D}\mathbf{U}^{-1} = \mathbf{L}'\mathbf{L}$. Choose a vector $\boldsymbol{u} \in \mathcal{N}[\mathbf{M}(\xi)]$. From the inequalities $0 = \boldsymbol{u}'\mathbf{M}(\xi)\boldsymbol{u} \geq \boldsymbol{u}'\mathbf{M}(\eta)\boldsymbol{u} \geq 0$ it follows that $(\mathbf{L}\boldsymbol{u})'(\mathbf{L}\boldsymbol{u}) = 0$, hence $\mathbf{L}\boldsymbol{u} = \boldsymbol{0}$. Therefore, $\mathbf{M}(\eta)\boldsymbol{u} = \boldsymbol{0}$. So we proved that $\mathcal{N}[\mathbf{M}(\xi)] \subset \mathcal{N}[\mathbf{M}(\eta)]$, and hence, according to Proposition II.5, $\mathcal{M}[\mathbf{M}(\eta)] \subset \mathcal{M}[\mathbf{M}(\xi)]$. Following the Proposition II.7, a functional g, which is estimable under the design η (i.e. $\boldsymbol{g} \in \mathcal{M}[\mathbf{M}(\eta)]$), is also estimable under the design ξ (i.e. $\boldsymbol{g} \in \mathcal{M}[\mathbf{M}(\xi)]$). Moreover, in such a case there are vectors $\boldsymbol{z}_g \in \mathcal{M}[\mathbf{M}(\xi)]$ and $\boldsymbol{s}_g \in \mathcal{M}[\mathbf{M}(\eta)]$ so that $\boldsymbol{g} = \mathbf{M}(\xi)\boldsymbol{z}_g = \mathbf{M}(\eta)\boldsymbol{s}_g$ and that $\text{var}_\xi g = \boldsymbol{z}_g'\mathbf{M}(\xi)\boldsymbol{z}_g$, $\text{var}_\eta g = \boldsymbol{s}_g'\mathbf{M}(\eta)\boldsymbol{s}_g$. We may write

$$\begin{aligned}
0 &\leq (\boldsymbol{z}_g - \boldsymbol{s}_g)'\mathbf{M}(\eta)(\boldsymbol{z}_g - \boldsymbol{s}_g)\\
&= \boldsymbol{z}_g'\mathbf{M}(\eta)\boldsymbol{z}_g + \boldsymbol{s}_g'\mathbf{M}(\eta)\boldsymbol{s}_g - 2\boldsymbol{z}_g'\mathbf{M}(\eta)\boldsymbol{s}_g\\
&\leq \boldsymbol{z}_g'\mathbf{M}(\xi)\boldsymbol{z}_g + \boldsymbol{s}_g'\mathbf{M}(\eta)\boldsymbol{s}_g - 2\boldsymbol{z}_g'\boldsymbol{g}\\
&= \boldsymbol{z}_g'\mathbf{M}(\xi)\boldsymbol{z}_g + \boldsymbol{s}_g'\mathbf{M}(\eta)\boldsymbol{s}_g - 2\boldsymbol{z}_g'\mathbf{M}(\xi)\boldsymbol{z}_g\\
&= \text{var}_\eta g - \text{var}_\xi g.
\end{aligned}$$

Thus $\xi \lesssim \eta$.

Suppose, in turn, that $\xi \lesssim \eta$. From the inequality $\text{var}_\xi g \leq \text{var}_\eta g$, assumed to hold for every functional g, it follows that $\boldsymbol{g} \notin \mathcal{M}[\mathbf{M}(\xi)]$ (i.e. $\text{var}_\xi g = \infty$) implies $\text{var}_\eta g = \infty$ (i.e. $\boldsymbol{g} \notin \mathcal{M}[\mathbf{M}(\eta)]$). It follows that $\mathcal{M}[\mathbf{M}(\eta)] \subset \mathcal{M}[\mathbf{M}(\xi)]$. Furthermore, for estimable g we obtain

$$\boldsymbol{g}'\mathbf{M}^+(\xi)\boldsymbol{g} \leq \boldsymbol{g}'\mathbf{M}^+(\eta)\boldsymbol{g}; \quad (\boldsymbol{g} \in \mathcal{M}[\mathbf{M}(\eta)]). \tag{1}$$

Take $\boldsymbol{u} \in \mathcal{M}[\mathbf{M}(\eta)]$ and denote $\boldsymbol{z} = \mathbf{M}(\xi)\boldsymbol{u}$, $\boldsymbol{s} = \mathbf{M}(\eta)\boldsymbol{u}$. From Eq. (1) we obtain

$$\begin{aligned} 0 &\leq (\boldsymbol{z}-\boldsymbol{s})'\mathbf{M}^{+}(\xi)(\boldsymbol{z}-\boldsymbol{s}) \\ &= \boldsymbol{z}'\mathbf{M}^{+}(\xi)\boldsymbol{z} + \boldsymbol{s}'\mathbf{M}^{+}(\xi)\boldsymbol{s} - 2\boldsymbol{s}'\mathbf{M}^{+}(\xi)\boldsymbol{z} \\ &= \leq \boldsymbol{z}'\mathbf{M}^{+}(\xi)\boldsymbol{z} + \boldsymbol{s}'\mathbf{M}^{+}(\eta)\boldsymbol{s} - 2\boldsymbol{s}'\boldsymbol{u} \quad \text{(Proposition II.5)} \\ &= \boldsymbol{z}'\mathbf{M}^{+}(\xi)\boldsymbol{z} - \boldsymbol{s}'\mathbf{M}^{+}(\eta)\boldsymbol{s} \\ &= \boldsymbol{u}'\mathbf{M}(\xi)\boldsymbol{u} - \boldsymbol{u}'\mathbf{M}(\eta)\boldsymbol{u}. \square \end{aligned}$$

COROLLARY. *If* $\mathbf{M}(\xi) \geq \mathbf{M}(\eta)$, *and* $\mathbf{M}(\xi) \neq \mathbf{M}(\eta)$, *then* $\xi < \eta$. *On the reverse, if* $\xi < \eta$, *then* $\mathcal{M}[\mathbf{M}(\eta)] \subset \mathcal{M}[\mathbf{M}(\xi)]$ *and*

$$\boldsymbol{u}'\mathbf{M}(\eta)\boldsymbol{u} \leq \boldsymbol{u}'\mathbf{M}(\xi)\boldsymbol{u}; \quad (\boldsymbol{u} \in \mathcal{M}[\mathbf{M}(\eta)]),$$

where there is a strict inequality at least for one vector $\boldsymbol{u} \in \mathcal{M}[\mathbf{M}(\eta)]$.

The proof follows from Propositions III.1 and III.2.

PROPOSITION III.3. *Suppose that* $\mathbf{M}(\xi)$, $\mathbf{M}(\eta)$ *are nonsingular. Then* $\xi \lesssim \eta$ *(respectively* $\xi < \eta$*) if and only if* $\mathbf{M}(\xi) \geq \mathbf{M}(\eta)$ *(respectively if and only if* $\mathbf{M}(\xi) \geq \mathbf{M}(\eta)$ *and* $\mathbf{M}(\xi) \neq \mathbf{M}(\eta)$*).*

The proof follows from Propositions III.2 and III.3.

EXERCISE III.1. Prove the statement: If for every $\vartheta \in \Theta$

$$\sum_{x \in X} \vartheta^2(x)\xi(x) \geq \sum_{x \in X} \vartheta^2(x)\eta(x),$$

then $\xi \lesssim \eta$.

HINT. $\vartheta(x) = \boldsymbol{\alpha}'\boldsymbol{f}(x) \Rightarrow \sum_{x \in X} \vartheta^2(x)\xi(x) = \boldsymbol{\alpha}'\mathbf{M}(\xi)\boldsymbol{\alpha}$.

EXERCISE III.2. Consider the regression model with $X = \langle a, b \rangle$, $f_1(x) = 1$, $f_2(x) = x$, ..., $f_m(x) = x^{m-1}$, $\sigma(x) = 1$ for every $x \in X$ (compare with Exercise II.6). Denote

$$v_i(\xi) \equiv \sum_{x \in X} x^i \xi(x).$$

(i) Verify that

$$\{\mathbf{M}(\xi)\}_{ij} = v_{i+j-2}(\xi); \quad (i, j = 1, ..., m).$$

(ii) Prove the following statement (cf. [2]): If $\mathbf{M}(\xi)$, $\mathbf{M}(\eta)$ are nonsingular, then

$$\xi \lesssim \eta \Leftrightarrow v_i(\xi) = v_i(\eta); \quad (i = 0, \ldots, 2m-3)$$
$$v_{2m-2}(\xi) \geq v_{2m-2}(\eta).$$

HINT. Denote $\Delta\mathbf{M} \equiv \mathbf{M}(\xi) - \mathbf{M}(\eta)$, $\Delta v_i \equiv v_i(\xi) - v_i(\eta)$. Obviously $\Delta v_0 = 0$. If $\Delta v_j \neq 0$ for some $1 \leq j \leq m-1$, take $c < 0$ and define the vector

$$\boldsymbol{u} = (c/\Delta v_j, 0, \ldots, 0, \quad u_j = 1, 0, \ldots, 0).$$

Prove that $c \mapsto -\infty$ implies $\boldsymbol{u}'\Delta\mathbf{M}\boldsymbol{u} \mapsto -\infty$, which is in contradiction with Proposition III.2. Hence $\Delta v_j = 0$ for $1 \leq j \leq m-1$. Proceed similarly for the other rows of the matrix $\Delta\mathbf{M}$.

Besides the uniform ordering of designs, also another ordering can be useful in experimental design (cf. [37]). For any information matrix $\mathbf{M}(\xi)$ denote the ordered eigenvalues of $\mathbf{M}(\xi)$ by $\lambda_1(\xi) \leq \lambda_2(\xi) \leq \ldots \leq \lambda_m(\xi)$. Multiple eigenvalues (i.e. eingenvalues corresponding to several orthonormal eigenvectors) appear in the sequence $\lambda_1(\xi), \ldots, \lambda_m(\xi)$ as many times as their multiplicity is large. We shall say that *according to Schur's ordering* a design ξ is better than the design η iff

$$\sum_{i=1}^{k} \lambda_i(\xi) \geq \sum_{i=1}^{k} \lambda_i(\eta); \quad (k = 1, \ldots, m)$$

and iff there is a strict inequality for one k at least. The designs ξ and η are said to be *equivalent according to Schur's ordering* iff there is the equality sign for $k \in \{1, \ldots, m\}$.

PROPOSITION III.4 (*Schur's theorem*) [111]. *If* $\mathbf{A}$ *is an* $m \times m$ *symmetric matrix with diagonal elements* $a_1, \ldots, a_m$ *and ordered eigenvalues* $\mu_1 \geq \ldots \geq \mu_m$, *then*

$$\sum_{i=1}^{k} a_i \leq \sum_{i=1}^{k} \mu_i; \quad (i = 1, \ldots, m-1),$$

$$\sum_{i=1}^{m} a_i = \sum_{i=1}^{m} \mu_i.$$

PROOF. Denote $\mathbf{D} \equiv \operatorname{diag}(\mu_1, \ldots, \mu_m)$. Let $\boldsymbol{u}_1, \ldots, \boldsymbol{u}_m$ be the orthonormal eigenvectors of $\mathbf{A}$ corresponding to $\mu_1, \ldots, \mu_m$, and denote $\mathbf{U} = (\boldsymbol{u}_1, \ldots, \boldsymbol{u}_m)$. Then $\mathbf{D} = \mathbf{U}'\mathbf{A}\mathbf{U}$, hence $\mathbf{A} = \mathbf{U}\mathbf{D}\mathbf{U}'$. It follows that

$$a_i = \sum_{j=1}^{m} (\{\mathbf{U}\}_{ij})^2 \mu_j.$$

Obviously,

$$\sum_{i=1}^{m} (\{\mathbf{U}\}_{ij})^2 = \sum_{j=1}^{m} (\{\mathbf{U}\}_{ij})^2 = 1.$$

Take $k \leqslant m-1$ and denote

$$t_j \equiv \sum_{i=1}^{k} (\{\mathbf{U}\}_{ij})^2 \leqslant 1.$$

Clearly, $\sum_{j=1}^{m} t_j = k$. We can write

$$\sum_{i=1}^{k} a_i = \sum_{j=1}^{m} t_j \mu_j,$$

hence

$$\sum_{i=1}^{k} a_i - \sum_{i=1}^{k} \mu_i = \sum_{j=1}^{m} t_j \mu_j - \sum_{j=1}^{k} \mu_j + \mu_k \left(k - \sum_{j=1}^{m} t_j\right)$$

$$= \sum_{j=1}^{k} (\mu_j - \mu_k)(t_j - 1) + \sum_{j=k+1}^{m} t_j(\mu_j - \mu_k) \leqslant 0.$$

Evidently,

$$\sum_{i=1}^{m} a_i = \sum_{j=1}^{m} \left(\sum_{i=1}^{m} (\{\mathbf{U}\}_{ij})^2 \right) \mu_j = \sum_{j=1}^{m} \mu_j. \square$$

We shall prove that Schur's ordering is a refinement of the uniform ordering. Indeed, we have the following.

PROPOSITION III.5. *If $\xi \sim \eta$, then ξ, η are equivalent in the sense of Schur's ordering. If $\xi < \eta$, then ξ is better in the sense of Schur's ordering than η.*

PROOF. The first statement is evident. To prove the second let $\mathbf{u}_1, \ldots, \mathbf{u}_m$ be the orthonormal eigenvectors of $\mathbf{M}(\xi)$ corresponding to $\lambda_1(\xi) \leqslant \ldots \leqslant \lambda_m(\xi)$. Then for every i

$$\lambda_i(\xi) = \mathbf{u}_i'\mathbf{M}(\xi)\mathbf{u}_i \geqslant \mathbf{u}_i'\mathbf{M}(\eta)\mathbf{u}_i \equiv h_i.$$

According to Proposition III.4

$$\sum_{i=1}^{k} h_i - \sum_{i=1}^{k} \lambda_i(\eta) = \left(\sum_{i=1}^{m} h_i - \sum_{i=1}^{m} \lambda_i(\eta)\right)$$

$$- \sum_{i=k+1}^{m} h_i + \sum_{i=k+1}^{m} \lambda_i(\eta) \geq 0.$$

Hence

$$\sum_{i=1}^{k} \lambda_i(\xi) \geq \sum_{i=1}^{k} h_i \geq \sum_{i=1}^{k} \lambda_i(\eta); \quad (k=1, \ldots, m),$$

i.e. ξ is better in the sense of Schur's ordering than η. □

For the connection of the Schur's ordering with optimality criteria see Proposition IV.30.

EXERCISE III.3. Take $s<m$. Let $\mathscr{H}$ be the subspace of R^m spanned by $\mathbf{e}^{(1)}, \ldots, \mathbf{e}^{(s)}$ (= the first s columns of the identity matrix). Define the partial uniform ordering as

$$\xi \underset{s}{\lesssim} \eta \Leftrightarrow \operatorname{var}_\xi g \leq \operatorname{var}_\eta g; \quad (\mathbf{g} \in \mathscr{H}).$$

Suppose that $\mathbf{M}(\xi)$ and $\mathbf{M}(\eta)$ are nonsingular.

(i) Define the partial information matrix of ξ by

$$\mathbf{M}^{(s)}(\xi) = [\mathbf{P}'\mathbf{M}^{-1}(\xi)\mathbf{P}]^{-1},$$

where

$$\mathbf{P} = (\mathbf{e}^{(1)}, \ldots, \mathbf{e}^{(s)}).$$

Prove that Proposition III.3 remains true when considering the partial ordering and taking $\mathbf{M}^{(s)}(\xi)$ and $\mathbf{M}^{(s)}(\eta)$ instead of $\mathbf{M}(\xi)$ and $\mathbf{M}(\eta)$.

(ii) Define the partial Schur's ordering of designs for the information matrices $\mathbf{M}^{(s)}(\xi)$, $\mathbf{M}^{(s)}(\eta)$. Prove Proposition III.5 for the partial orderings.

HINT. Follow the proofs of Propositions III.2—III.5.

DEFINITION III.2. *A design ξ is said to be admissible if there is no design η such that $\eta < \xi$.*

PROPOSITION III.6 [26]. *Suppose that $\mathbf{M}(\xi)$ is nonsingular. If there is a positive definite matrix $\mathbf{V}$ such that*

$$\mathbf{f}'(x)\mathbf{V}\mathbf{f}(x) \leq 1; \quad (x \in X),$$

$$\mathbf{f}'(x)\mathbf{V}\mathbf{f}(x) = 1; \quad (x \in X_\xi),$$

then ξ is an admissible design.

PROOF. Let $\mathbf{v}_1, \ldots, \mathbf{v}_m$ be orthonormal eigenvectors of $\mathbf{V}$. Then $\mathbf{D} \equiv (\mathbf{v}_1, \ldots, \mathbf{v}_m)'\mathbf{V}(\mathbf{v}_1, \ldots, \mathbf{v}_m) = \operatorname{diag}(d_1, \ldots, d_m)$ with $d_i \geqslant 0$. Hence

$$\mathbf{V} = (\mathbf{v}_1, \ldots, \mathbf{v}_m)\mathbf{D}(\mathbf{v}_1, \ldots, \mathbf{v}_m)' = \sum_{i=1}^{m} \sqrt{d_i}\mathbf{v}_i(\sqrt{d_i}\mathbf{v}_i)'.$$

Thus the assumptions in the proposition can be rewritten as

$$\begin{aligned} &\sum_{i=1}^{m} (\mathbf{f}'(x)\mathbf{v}_i)^2 d_i \leqslant 1; \quad (x \in X), \\ &\sum_{i=1}^{m} (\mathbf{f}'(x)\mathbf{v}_i)^2 d_i = 1; \quad (x \in X_\xi). \end{aligned} \tag{2}$$

If $\eta \lesssim \xi$, then it follows from the second part of Proposition III.2 that

$$\mathbf{u}'\mathbf{M}(\eta)\mathbf{u} \geqslant \mathbf{u}'\mathbf{M}(\xi)\mathbf{u}; \quad (\mathbf{u} \in R^m). \tag{3}$$

Particularly, setting $\mathbf{v}_i$ instead of $\mathbf{u}$ we obtain

$$\mathbf{v}_i'\mathbf{M}(\eta)\mathbf{v}_i - \mathbf{v}_i'\mathbf{M}(\xi)\mathbf{v}_i \geqslant 0; \quad (i = 1, \ldots, m).$$

On the other hand, from Eq. (2) it follows that

$$\begin{aligned} &\sum_{i=1}^{m} [\mathbf{v}_i'\mathbf{M}(\eta)\mathbf{v}_i - \mathbf{v}_i'\mathbf{M}(\xi)\mathbf{v}_i]d_i \\ &= \sum_{x \in X} \sum_{i=1}^{m} (\mathbf{f}'(x)\mathbf{v}_i)^2 d_i \eta(x) - \sum_{x \in X} \sum_{i=1}^{m} (\mathbf{f}'(x)\mathbf{v}_i)^2 d_i \xi(x) \leqslant 0, \end{aligned}$$

which is possible only if

$$\mathbf{v}_i'\mathbf{M}(\eta)\mathbf{v}_i = \mathbf{v}_i'\mathbf{M}(\xi)\mathbf{v}_i; \quad (i = 1, \ldots, m).$$

Hence it follows from Eq. (3) that

$$\begin{aligned} 0 &\leqslant (\mathbf{v}_i + \mathbf{v}_j)'\mathbf{M}(\eta)(\mathbf{v}_i + \mathbf{v}_j) - (\mathbf{v}_i + \mathbf{v}_j)'\mathbf{M}(\xi)(\mathbf{v}_i + \mathbf{v}_j) \\ &= 2[\mathbf{v}_i'\mathbf{M}(\eta)\mathbf{v}_j - \mathbf{v}_i'\mathbf{M}(\xi)\mathbf{v}_j], \end{aligned}$$

and similarly

$$\begin{aligned} 0 &\leqslant (\mathbf{v}_i - \mathbf{v}_j)'\mathbf{M}(\eta)(\mathbf{v}_i - \mathbf{v}_j) - (\mathbf{v}_i - \mathbf{v}_j)'\mathbf{M}(\xi)(\mathbf{v}_i - \mathbf{v}_j) \\ &= -2[\mathbf{v}_i'\mathbf{M}(\eta)\mathbf{v}_j - \mathbf{v}_i'\mathbf{M}(\xi)\mathbf{v}_j]. \end{aligned}$$

Thus

$$\mathbf{v}_i'\mathbf{M}(\eta)\mathbf{v}_j = \mathbf{v}_i'\mathbf{M}(\xi)\mathbf{v}_j\,; \quad (i, j=1, \ldots, m)$$

and therefore

$$\mathbf{M}(\xi) = \mathbf{M}(\eta),$$

i.e. $\xi \sim \eta$. So there is no design η such that $\eta < \xi$. □

COROLLARY. *Let* $\mathbf{u}_1, \ldots, \mathbf{u}_m$ *be arbitrary, linearly independent vectors taken from* R^m. *Denote*

$$Z \equiv \left\{ \bar{x} \in X\colon \sum_{i=1}^{m} [\mathbf{f}'(\bar{x})\mathbf{u}_i]^2 = \max_{x \in X} \sum_{i=1}^{m} [\mathbf{f}'(x)\mathbf{u}_i]^2 \right\}.$$

Then any design ξ *which is supported by* Z *(i.e.* $\{x\colon \xi(x)>0\} = Z$*) is an admissible design.*

PROOF. Take $\mathbf{U} = (\mathbf{u}_1, \ldots, \mathbf{u}_m)$. The matrix $\mathbf{UU}'$ is positive definite because $\mathbf{a}'\mathbf{UU}'\mathbf{a} = \|\mathbf{U}'\mathbf{a}\|^2 \geqslant 0$ for any $\mathbf{0} \neq \mathbf{a} \in R^m$ and because $\mathbf{U}$ is nonsingular. Take

$$\mathbf{V} = \mathbf{UU}' / \max_{x \in X} \sum_{i=1}^{m} [\mathbf{f}'(x)\mathbf{u}_i]^2$$

and use Proposition III.6. □

Denote the smallest convex set in R^m which contains the set $T \equiv \{\mathbf{f}(x)\colon x \in X\} \cup \{-\mathbf{f}(x)\colon x \in X\}$ by S. (S is the convex hull of T.) This set can be expressed as

$$S = \left\{ \sum_{x \in X'} \mathbf{f}(x)\xi(x) - \sum_{x \in X - X'} \mathbf{f}(x)\xi(x)\colon X' \subset X,\ \xi \in \Xi \right\}. \quad (4)$$

The set S is important for the comparison of designs of experiments.

PROPOSITION III.7 [8]. *To every design* ξ *there is a design* η *such that* $\eta \lesssim \xi$ *and that* $\{\mathbf{f}(x)\colon \eta(x)>0\}$ *is a subset of the boundary of* S.

PROOF. Suppose that $\xi(x_0) > 0$ and that $\mathbf{f}(x_0)$ is an inner point of S. Then there is $c > 1$ such that $c\mathbf{f}(x_0)$ is on the boundary of S. From the definition of S it follows that there are vectors $\mathbf{f}(x_1), \ldots, \mathbf{f}(x_r)$ from the boundary of S, and numbers $\gamma_1, \ldots, \gamma_r \in \langle -1, 1 \rangle$ such that $\sum_{i=1}^{r} |\gamma_i| = 1$

and that

$$c\boldsymbol{f}(x_0)=\sum_{i=1}^{r}\gamma_i\boldsymbol{f}(x_i).$$

Define the design η:

$$\begin{aligned}\eta(x)&=0 && \text{if } x=x_0,\\ &=\xi(x_i)+\xi(x_0)|\gamma_i| && \text{if } x=x_i,\\ &=\xi(x) && \text{if } x\notin\{x_0, x_1, \ldots, x_r\}.\end{aligned}$$

We can write

$$\begin{aligned}&\mathbf{M}(\eta)-\mathbf{M}(\xi)\\ &=\xi(x_0)\sum_{i=1}^{r}\boldsymbol{f}(x_i)\boldsymbol{f}'(x_i)|\gamma_i|-\boldsymbol{f}(x_0)\boldsymbol{f}'(x_0)\xi(x_0),\end{aligned}$$

hence for every vector $\boldsymbol{u}\in R^m$

$$\begin{aligned}&\boldsymbol{u}'[\mathbf{M}(\eta)-\mathbf{M}(\xi)]\boldsymbol{u}\\ &=\xi(x_0)\left\{\sum_{i=1}^{r}[\boldsymbol{u}'\boldsymbol{f}(x_i)]^2|\gamma_i|-\frac{1}{c^2}\left[\sum_{i=1}^{r}\boldsymbol{u}'\boldsymbol{f}(x_i)\gamma_i\right]^2\right\}\\ &\geqslant\xi(x_0)\left\{\sum_{i=1}^{r}[\boldsymbol{u}'\boldsymbol{f}(x_i)]^2|\gamma_i|-\left[\sum_{i=1}^{r}\boldsymbol{u}'\boldsymbol{f}(x_i)\gamma_i\right]^2\right\}\geqslant 0,\end{aligned}$$

since, using the Schwarz inequality, we obtain

$$\begin{aligned}&\left[\sum_{i=1}^{r}\boldsymbol{u}'\boldsymbol{f}(x_i)\gamma_i\right]^2=\left[\sum_{i=1}^{r}\sqrt{|\gamma_i|}\frac{\gamma_i}{\sqrt{|\gamma_i|}}\boldsymbol{u}'\boldsymbol{f}(x_i)\right]^2\\ &\leqslant\sum_{i=1}^{r}|\gamma_i|\sum_{i=1}^{r}|\gamma_i|[\boldsymbol{u}'\boldsymbol{f}(x_i)]^2=\sum_{i=1}^{r}|\gamma_i|[\boldsymbol{u}'\boldsymbol{f}(x_i)]^2.\end{aligned}$$

Therefore, according to Proposition III.2 we have $\eta\lesssim\xi$. □

PROPOSITION III.8 (*the Caratheodory theorem*). *If T is a subset of the k-dimensional Euclidean space, then every point of the convex hull of T*

$$\operatorname{co}(T)\equiv\left\{\boldsymbol{z}\colon \boldsymbol{z}=\sum_{i=1}^{n(z)}\beta_i\boldsymbol{t}_i,\ \beta_i\in\langle 0, 1\rangle,\ \sum_{i=1}^{n(z)}\beta_i=1,\ \boldsymbol{t}_i\in T\right\}$$

can be expressed as an affine combination of at most $k+1$ points from the set T. That means to every $\boldsymbol{q}\in \mathrm{co}\,(T)$ there are points

$$\boldsymbol{t}_1, \ldots, \boldsymbol{t}_{k+1}\in T$$

and numbers

$$\gamma_1, \ldots, \gamma_{k+1}\in\langle 0, 1\rangle, \sum_{i=1}^{k+1} \gamma_i = 1$$

such that

$$\boldsymbol{q} = \sum_{i=1}^{k+1} \gamma_i \boldsymbol{t}_i.$$

PROOF. Suppose that $\boldsymbol{q}=\sum_{i=1}^{r} \beta_i \boldsymbol{t}_i$ for some $\beta_i\in(0, 1)$, $\sum_{i=1}^{r} \beta_i = 1$, $\boldsymbol{t}_i\in T$, $r\geqslant k+2$. In the set $\{\boldsymbol{t}_1, \ldots, \boldsymbol{t}_{k+2}\}\subset R^k$ there are at least two vectors which are linear combinations of the other vectors. It follows that

$$\sum_{i=1}^{k+2} \delta_i' \boldsymbol{t}_i = 0, \quad \sum_{i=1}^{k+2} \delta_i'' \boldsymbol{t}_i = 0$$

for some δ_i', δ_i''; $(i=1, \ldots, k+2)$, $\sum_{i=1}^{k+2} \delta_i'^2>0$, $\sum_{i=1}^{k+2} \delta_i''^2>0$.

If $\sum_{i=1}^{k+2} \delta_i' = \sum_{i=1}^{k+2} \delta_i'' = 0$, then take an arbitrary number $c\in R$.

If $\sum_{i=1}^{k+2} \delta_i'' \neq 0$, then take

$$c = -\left(\sum_{i=1}^{k+2} \delta_i'\right) \Big/ \left(\sum_{i=1}^{k+2} \delta_i''\right).$$

Define

$$\begin{aligned} \delta_i &= \delta_i' + c\delta_i'' && \text{if } 1\leqslant i\leqslant k+2,\\ &= 0 && \text{if } k+2<i\leqslant r. \end{aligned}$$

We can write for every $\varepsilon\in R$

$$\sum_{i=1}^{r} (\beta_i + \varepsilon\delta_i) = 1 + \varepsilon\left(\sum_{i=1}^{k+2} \delta_i' + c\sum_{i=1}^{k+2} \delta_i''\right) = 1,$$

$$\sum_{i=1}^{r} (\beta_i + \varepsilon\delta_i)\boldsymbol{t}_i = \boldsymbol{q} + \varepsilon \left(\sum_{i=1}^{k+2} \delta'_i \boldsymbol{t}_i + c \sum_{i=1}^{k+2} \delta''_i \boldsymbol{t}_i \right) = \boldsymbol{q}. \tag{5}$$

Choose the subscript j according to

$$\left|\frac{\beta_j}{\delta_j}\right| = \min \left\{ \left|\frac{\beta_i}{\delta_i}\right| : 1 \leqslant i \leqslant k+2,\ \delta_i \neq 0 \right\} \tag{6}$$

and take

$$\varepsilon = -\frac{\beta_j}{\delta_j}. \tag{7}$$

Define

$$\gamma_i = \beta_i + \varepsilon\delta_i; \quad (i = 1, \ldots, r).$$

According to Eq. (5) we have

$$\sum_{i=1}^{r} \gamma_i = 1, \quad \sum_{i=1}^{r} \gamma_i \boldsymbol{t}_i = \boldsymbol{q}.$$

From Eqs. (6) and (7) we obtain $\gamma_i \geqslant 0$; $(i = 1, \ldots, r)$, $\gamma_j = 0$. Thus there are only $r-1$ nonzero coefficients in the sum $\sum_{i=1}^{r} \gamma_i \boldsymbol{t}_i = \boldsymbol{q}$.

We can continue to eliminate the coefficients until reaching the number $k+1$ of nonzero coefficients. □

We shall use further that any continuous mapping maps compact sets onto compact sets (cf. e.g. [99], Second part, § 6.2).

PROPOSITION III.9. *If T is a compact subset of R^k, then* co (T) *is compact in R^k.*

PROOF. Denote the k-th Cartesian power of the set T by T^k (similarly for $\langle 0, 1\rangle^k$). Define a mapping

$$\varphi: T^{k+1} \times \langle 0, 1\rangle^{k+1} \mapsto R^k$$

by the expression

$$\varphi(\boldsymbol{t}_1, \ldots, \boldsymbol{t}_{k+1}, \gamma_1, \ldots, \gamma_{k+1}) = \sum_{i=1}^{k+1} \gamma_i \boldsymbol{t}_i.$$

The mapping φ is continuous and it is defined on a compact set (since a Cartesian product of compact sets is compact, cf. [98], p. 153). From Proposition III.8 it follows that the range of the mapping φ is the compact set co (T). □

PROPOSITION III.10. *The set of all information matrices*

$$\mathfrak{M} = \{\mathbf{M}(\xi) : \xi \in \Xi\}$$

is a compact set.

PROOF. Denote $\mathscr{C} \equiv \{\boldsymbol{f}(x)\boldsymbol{f}'(x) : x \in X\}$. The set $\mathscr{C}$ is the image of the compact set X under the mapping $x \in X \mapsto \boldsymbol{f}(x)\boldsymbol{f}'(x)$, hence it is compact. From

$$\mathbf{M}(\xi) = \sum_{x \in X} \boldsymbol{f}(x)\boldsymbol{f}'(x)\xi(x)$$

it follows that $\mathfrak{M} = \text{co}\,(\mathscr{C})$, thus $\mathfrak{M}$ is a compact set. □

PROPOSITION III.11 [26]. *To every design ξ there is an equivalent design η such that its support $X_\eta = \{x : \eta(x) > 0\}$ has at most $m(m+1)/2+1$ elements.*

PROOF. Denote the $[m(m+1)/2]$-dimensional vector having the components $f_i(x)f_j(x)$; $(1 \leq i \leq j \leq m)$ by $\boldsymbol{a}(x)$. With every vector belonging to the convex set

$$\text{co}\,(\{\boldsymbol{a}(x) : x \in X\})$$

we can associate exactly one information matrix such that the components of the vector are the elements of the matrix which are on and above the diagonal. From Proposition III.8 it follows that to every design ξ there are numbers $\gamma_s \in \langle 0, 1 \rangle$ and points $x_s \in X$; $(s = 1, \ldots, [m(m+1)/2+1])$ such that $\sum_s \gamma_s = 1$ and

$$\sum_s \gamma_s f_i(x_s) f_j(x_s) = \{\mathbf{M}(\xi)\}_{ij}; \quad (1 \leq i \leq j \leq m).$$

Define $\eta(x_s)=\gamma_s$; $(s=1, \ldots, [m(m+1)/2+1])$. From the symmetry of information matrices it follows that $\mathbf{M}(\eta)=\mathbf{M}(\xi)$. □

III.3. PROPERTIES OF THE FUNCTION $\mathbf{M}(\xi)\in\mathfrak{M}\mapsto \mathrm{var}_\xi g$

In this chapter g is an arbitrary but fixed linear functional defined on Θ. We shall discuss how do changes of the design ξ affect the variance $\mathrm{var}_\xi g$. Since $\mathrm{var}_\xi g$ depends on ξ through $\mathbf{M}(\xi)$, the function $\mathbf{M}(\xi)\in\mathfrak{M}\mapsto \mathrm{var}_\xi g$ will be investigated.

PROPOSITION III.12. *Let* $\mathbf{A}$, $\mathbf{B}$ *be symmetric* $m\times m$ *matrices and let* $\mathbf{A}$ *be positive definite. Then there is a nonsingular* $m\times m$ *matrix* $\mathbf{U}$ *such that*

$$\mathbf{UAU}'=\mathbf{I},$$

$$\mathbf{UBU}'=\Lambda,$$

where Λ *is a diagonal matrix.*

PROOF. Let $\mathbf{w}_1, \ldots, \mathbf{w}_m$ be orthonormal eigenvalues of the matrix $\mathbf{A}$ and denote $\mathbf{W}\equiv(\mathbf{w}_1, \ldots, \mathbf{w}_m)$. The matrix $\mathbf{W}'\mathbf{AW}\equiv\mathbf{D}$ is a diagonal matrix with positive diagonal elements. Define $\mathbf{C}\equiv\mathbf{W}\sqrt{\mathbf{D}}\mathbf{W}'$. Obviously $\mathbf{C}$ is nonsingular. We have

$$\mathbf{A}=\mathbf{C}^2.$$

Define

$$\mathbf{S}\equiv\mathbf{C}^{-1}\mathbf{BC}^{-1}.$$

The matrix $\mathbf{S}$ being symmetric, there are m orthonormal eigenvectors of $\mathbf{S}$, say $\mathbf{s}_1, \ldots, \mathbf{s}_m$, corresponding to eigenvalues $\lambda_1, \ldots, \lambda_m$. Define

$$\mathbf{z}_i\equiv\mathbf{C}^{-1}\mathbf{s}_i; \quad (i=1, \ldots, m),$$

and

$$\mathbf{U}'\equiv(\mathbf{z}_1, \ldots, \mathbf{z}_m).$$

We have

$$\mathbf{z}_i'\mathbf{Az}_k=(\mathbf{Cz}_i)'(\mathbf{Cz}_k)=\mathbf{s}_i'\mathbf{s}_k,$$

and

$$\mathbf{z}_i'\mathbf{B}\mathbf{z}_k = \mathbf{z}_i'\mathbf{C}\mathbf{S}\mathbf{C}\mathbf{z}_k = \mathbf{s}_i'\mathbf{s}_k\lambda_k.$$

It follows that

$$\mathbf{U}\mathbf{A}\mathbf{U}' = \mathbf{I}, \quad \mathbf{U}\mathbf{B}\mathbf{U}' = \mathbf{\Lambda},$$

where $\mathbf{\Lambda} = \operatorname{diag}(\lambda_1, \ldots, \lambda_m)$. □

PROPOSITION III.13. *The function*

$$\mathbf{M}(\xi) \in \mathfrak{M} \mapsto \operatorname{var}_\xi g$$

is convex on the set of all information matrices $\mathfrak{M}$, *and it is strictly convex on the set of all nonsingular information matrices.*

PROOF. We have to prove that

$$\operatorname{var}_{(1-\beta)\xi+\beta\eta} g \leq (1-\beta) \operatorname{var}_\xi g + \beta \operatorname{var}_\eta g$$

for every $\beta \in \langle 0, 1 \rangle$, $\xi, \eta \in \Xi$. This inequality holds obviously if $\operatorname{var}_\xi g = \infty$ or if $\operatorname{var}_\eta g = \infty$. Suppose therefore that $\mathbf{M}(\xi)$, $\mathbf{M}(\eta)$ are nonsingular. Then, according to Proposition III.12, there is a nonsingular matrix $\mathbf{U}$ such that $\mathbf{U}'\mathbf{M}(\xi)\mathbf{U} = \mathbf{I}$, $\mathbf{U}'\mathbf{M}(\eta)\mathbf{U} = \mathbf{\Lambda} \equiv \operatorname{diag}(\lambda_1, \ldots, \lambda_m)$. From the strict convexity of the function $x \in (0, \infty) \mapsto x^{-1}$ it follows that

$$\begin{aligned}
&\mathbf{g}'[(1-\beta)\mathbf{M}(\xi) + \beta\mathbf{M}(\eta)]^{-1}\mathbf{g} \\
&= \mathbf{g}'\mathbf{U}[(1-\beta)\mathbf{I} + \beta\mathbf{\Lambda}]^{-1}\mathbf{U}'\mathbf{g} \\
&= \mathbf{g}'\mathbf{U} \operatorname{diag}([(1-\beta)+\beta\lambda_1]^{-1}, \ldots, [(1-\beta)+\beta\lambda_m]^{-1})\mathbf{U}'\mathbf{g} \\
&\leq \mathbf{g}'\mathbf{U}[(1-\beta)\mathbf{I} + \beta\mathbf{\Lambda}^{-1}]\mathbf{U}'\mathbf{g} \\
&= (1-\beta)\mathbf{g}'\mathbf{M}^{-1}(\xi)\mathbf{g} + \mathbf{g}'\mathbf{M}^{-1}(\eta)\mathbf{g}.
\end{aligned}$$

The equality sing is achieved exactly if $\lambda_1 = \ldots = \lambda_m = 1$, that is if $\mathbf{M}(\xi) = \mathbf{M}(\eta)$.

In case that $\det \mathbf{M}(\xi) = 0$ or $\det \mathbf{M}(\eta) = 0$ but

$$\begin{aligned}
\operatorname{var}_\xi g &= \mathbf{g}'\mathbf{M}^{-}(\xi)\mathbf{g} < \infty, \\
\operatorname{var}_\eta g &= \mathbf{g}'\mathbf{M}^{-}(\eta)\mathbf{g} < \infty,
\end{aligned}$$

we can prove as above that

$$\mathbf{g}'[(1-\beta)\mathbf{M}(\xi) + \beta\mathbf{M}(\eta) + \varepsilon\mathbf{I}]^{-1}\mathbf{g}$$

$$\leqslant(1-\beta)\boldsymbol{g}'[\mathbf{M}(\xi)+\varepsilon\mathbf{I}]^{-1}\boldsymbol{g}+\beta\boldsymbol{g}'[\mathbf{M}(\eta)+\varepsilon\mathbf{I}]^{-1}\boldsymbol{g}.$$

Taking the limit $\varepsilon\searrow 0$ we obtain the needed result since, according to Proposition II.9, $\lim_{\varepsilon\searrow 0}(\mathbf{M}(\eta)+\varepsilon\mathbf{I})^{-1}$ and $\lim_{\varepsilon\searrow 0}(\mathbf{M}(\xi)+\varepsilon\mathbf{I})^{-1}$ are g-inverses of $\mathbf{M}(\eta)$ and $\mathbf{M}(\xi)$. □

An arbitrary real-valued function φ defined on a topological space Z is said to be *lower semicontinuous* at the point $z_0\in Z$ if for every $\varepsilon>0$ there is a neighbourhood U of z_0 such that $z\in U$ implies $\varphi(z)>\varphi(z_0)-\varepsilon$.

If Z is a metric space, then the function φ is lower semicontinuous at z_0 if and only if

$$\liminf_{n\to\infty}\varphi(z_n)\geqslant\varphi(z_0)$$

for every sequence $\{z_n\}_{n=1}^{\infty}$ of points from the set Z converging to z_0 (cf. [99], Second part, § 6.3).

PROPOSITION III.14. *The function*

$$M(\xi)\in\mathfrak{M}\mapsto \mathrm{var}_\xi g$$

is lower semicontinuous. It is continuous at $\mathbf{M}(\xi)\in\mathfrak{M}$ *if* $\det\mathbf{M}(\xi)\neq 0$ *or if* $\mathrm{var}_\xi g=\infty$.

PROOF. Denote $\mathbf{M}\equiv\mathbf{M}(\xi)$, $\mathbf{M}_n\equiv\mathbf{M}(\xi_n)$. Suppose that

$$\lim_{n\to\infty}\mathbf{M}_n=\mathbf{M}. \tag{8}$$

In the first part of the proof suppose that $\boldsymbol{g}\in\mathcal{M}(\mathbf{M})$. Denote the dimension of the linear space $\mathcal{M}(\mathbf{M})$ by r. Let $\boldsymbol{u}_1, \ldots, \boldsymbol{u}_r$ be an orthonormal basis in $\mathcal{M}(\mathbf{M})$. Denote

$$\boldsymbol{u}_0=-\frac{1}{2}\sum_{i=1}^{r}\boldsymbol{u}_i$$

and denote by V the simplex in $\mathcal{M}(\mathbf{M})$ having the vertices $\boldsymbol{u}_0, \boldsymbol{u}_1, \ldots, \boldsymbol{u}_r$. The point $\boldsymbol{0}$ is an inner point of the simplex V. Therefore, to every vector $\boldsymbol{z} \in \mathcal{M}(\mathbf{M})$ there is a number $c>0$ such that $c\boldsymbol{z}$ is on the boundary $\hat{V}$ of the simplex V,

$$\hat{V} \equiv \left\{ \sum_{i=0}^{r} \lambda_i \boldsymbol{u}_i : \sum_{i=0}^{r} \lambda_i = 1,\ \lambda_0, \ldots, \lambda_r \in \langle 0, 1 \rangle, \exists_j \lambda_j = 0 \right\}.$$

Using Eq. (24) in Chapter II we can write

$$\begin{aligned}
\mathrm{var}_{\xi} g &= \sup \left\{ \frac{(\boldsymbol{g}'\boldsymbol{\alpha})^2}{\boldsymbol{\alpha}'\mathbf{M}\boldsymbol{\alpha}} : \boldsymbol{\alpha} \in \mathcal{M}(\mathbf{M}),\ \boldsymbol{\alpha} \neq \boldsymbol{0} \right\} \\
&= \sup \left\{ \frac{(\boldsymbol{g}'\boldsymbol{\alpha})^2}{\boldsymbol{\alpha}'\mathbf{M}\boldsymbol{\alpha}} : \boldsymbol{\alpha} \in V,\ \boldsymbol{\alpha} \neq \boldsymbol{0} \right\} \\
&= \sup \left\{ \frac{(\boldsymbol{g}'\boldsymbol{\alpha})^2}{\boldsymbol{\alpha}'\mathbf{M}\boldsymbol{\alpha}} : \boldsymbol{\alpha} \in \hat{V} \right\}.
\end{aligned}$$

We shall prove first that there is n_0 such that for $n \geq n_0$

$$\mathcal{M}(\mathbf{M}) \cap \mathcal{N}(\mathbf{M}_n) = \{\boldsymbol{0}\}. \tag{9}$$

We have

$$d \equiv \min \{\boldsymbol{\alpha}'\mathbf{M}\boldsymbol{\alpha} : \boldsymbol{\alpha} \in \hat{V}\} > 0, \tag{10}$$

since $\hat{V}$ is compact and $\boldsymbol{0} \notin \hat{V}$. From Eq. (8) it follows that for some n_0 and for every $n \geq n_0$

$$\sup \{|\boldsymbol{u}_i'(\mathbf{M}_n - \mathbf{M})\boldsymbol{u}_j| : i, j = 0, \ldots, r\} < \frac{d}{2}.$$

Therefore,

$$\boldsymbol{u}_i'\mathbf{M}_n\boldsymbol{u}_j > \boldsymbol{u}_i'\mathbf{M}\boldsymbol{u}_j - \frac{d}{2};\ (n \geq n_0,\ i, j = 0, \ldots, r).$$

It follows that for every $\boldsymbol{\alpha} = \sum_i \lambda_i \boldsymbol{u}_i \in \hat{V}$

$$\boldsymbol{\alpha}'\mathbf{M}_n\boldsymbol{\alpha} = \sum_{i,j} \lambda_i \boldsymbol{u}_i'\mathbf{M}_n\boldsymbol{u}_j\lambda_j > \boldsymbol{\alpha}'\mathbf{M}\boldsymbol{\alpha} - \frac{d}{2} > \frac{d}{2}; \tag{11}$$

$(n \geq n_0)$.

If $\boldsymbol{0} \neq \boldsymbol{z} \in \mathcal{M}(\mathbf{M}) \cap \mathcal{N}(\mathbf{M}_n)$, then $\beta\boldsymbol{z} \in \hat{V}$ for some $\beta > 0$, and $\beta^2\boldsymbol{z}'\mathbf{M}_n\boldsymbol{z} = 0$.

This is in contradiction with the inequality (11). In this way, Eq. (9) has been proven.

When $\boldsymbol{g}\in\mathcal{M}(\mathbf{M}_n)$, Eq. (9) implies

$$\begin{aligned}\mathrm{var}_{\xi_n} g &= \sup\left\{\frac{(\mathbf{g}'\boldsymbol{\alpha})^2}{\boldsymbol{\alpha}'\mathbf{M}_n\boldsymbol{\alpha}} : \boldsymbol{\alpha}\in R^m - \mathcal{N}(\mathbf{M}_n)\right\} \\ &\geqslant \sup\left\{\frac{(\mathbf{g}'\boldsymbol{\alpha})^2}{\boldsymbol{\alpha}'\mathbf{M}_n\boldsymbol{\alpha}} : \boldsymbol{\alpha}\in\mathcal{M}(\mathbf{M}),\ \boldsymbol{\alpha}\neq \mathbf{0}\right\} \\ &= \sup\left\{\frac{(\mathbf{g}'\boldsymbol{\alpha})^2}{\boldsymbol{\alpha}'\mathbf{M}_n\boldsymbol{\alpha}} : \boldsymbol{\alpha}\in\hat{V}\right\};\quad (n\geqslant n_0).\end{aligned} \tag{12}$$

When in turn $\boldsymbol{g}\notin\mathcal{M}(\mathbf{M}_n)$, the inequality in Eq. (12) is evident because $\mathrm{var}_{\xi_n} g=\infty$. Using Eqs. (10)—(12) we obtain for $n\geqslant n_0$

$$\begin{aligned}&\mathrm{var}_{\xi} g - \mathrm{var}_{\xi_n} g \leqslant \sup\left\{\frac{(\mathbf{g}'\boldsymbol{\alpha})^2}{\boldsymbol{\alpha}'\mathbf{M}\boldsymbol{\alpha}} : \boldsymbol{\alpha}\in\hat{V}\right\} - \sup\left\{\frac{(\mathbf{g}'\boldsymbol{\alpha})^2}{\boldsymbol{\alpha}'\mathbf{M}_n\boldsymbol{\alpha}} : \boldsymbol{\alpha}\in\hat{V}\right\} \\ &\leqslant \sup\left\{\frac{(\mathbf{g}'\boldsymbol{\alpha})^2}{\boldsymbol{\alpha}'\mathbf{M}\boldsymbol{\alpha}\boldsymbol{\alpha}'\mathbf{M}_n\boldsymbol{\alpha}}\left[\boldsymbol{\alpha}'\mathbf{M}_n\boldsymbol{\alpha} - \boldsymbol{\alpha}'\mathbf{M}\boldsymbol{\alpha}\right] : \boldsymbol{\alpha}\in\hat{V}\right\} \\ &\leqslant \frac{4\|\boldsymbol{g}\|^2}{d^2}\max\{\|\boldsymbol{u}_i\|^2 : 0\leqslant i\leqslant r\}\max\{|\boldsymbol{u}_i'(\mathbf{M}_n-\mathbf{M})\boldsymbol{u}_j| : i, j=0, \dots, r\}.\end{aligned}$$

The last maximum converges to zero when $n\to\infty$. Therefore,

$$\limsup_{n\to\infty}\left[\mathrm{var}_{\xi} g - \mathrm{var}_{\xi_n} g\right]\leqslant 0,$$

that means

$$\liminf_{n\to\infty}\mathrm{var}_{\xi_n} g \geqslant \mathrm{var}_{\xi} g,$$

which is the needed lower semicontinuity. In the particular case of $\det \mathbf{M}(\xi)\neq 0$, it follows from Eq. (9) that $\mathcal{M}(\mathbf{M}_n)=\mathcal{M}(\mathbf{M})=R^m$. In this case it can be proved as above that

$$\begin{aligned}&|\mathrm{var}_{\xi_n} g - \mathrm{var}_{\xi} g| \leqslant \sup\left\{\left|\frac{(\mathbf{g}'\boldsymbol{\alpha})^2}{\boldsymbol{\alpha}'\mathbf{M}_n\boldsymbol{\alpha}} - \frac{(\mathbf{g}'\boldsymbol{\alpha})^2}{\boldsymbol{\alpha}'\mathbf{M}\boldsymbol{\alpha}}\right| : \boldsymbol{\alpha}\in\hat{V}\right\} \\ &\leqslant \frac{4\|\boldsymbol{g}\|^2}{d^2}\max\{\|\boldsymbol{u}_i\|^2 : 0\leqslant i\leqslant r\}\max\{|\boldsymbol{u}_i'(\mathbf{M}_n-\mathbf{M})\boldsymbol{u}_j| : i, j=0, \dots, r\}.\end{aligned}$$

Thus

$$\lim_{n\to\infty} \mathrm{var}_{\xi_n} g = \mathrm{var}_\xi g .$$

Finally, suppose that $\mathrm{var}_\xi g = \infty$, i.e. $\boldsymbol{g} \notin \mathcal{M}(\mathbf{M})$. According to Proposition II.5, we can decompose the vector $\boldsymbol{g}$ as $\boldsymbol{g} = \boldsymbol{g}_1 + \boldsymbol{g}_2$ with $\boldsymbol{g}_1 \in \mathcal{M}(\mathbf{M})$, $0 \neq \boldsymbol{g}_2 \in \mathcal{N}(\mathbf{M})$. In case of $\mathrm{var}_{\xi_n} g < \infty$ for some n, according to Proposition II.2, there is $\boldsymbol{z}_n \in \mathcal{M}(\mathbf{M}_n)$ such that $\boldsymbol{g} = \mathbf{M}_n \boldsymbol{z}_n$ and that $\mathrm{var}_{\xi_n} g = \boldsymbol{z}_n' \mathbf{M}_n \boldsymbol{z}_n$. Using the Schwarz inequality we can write

$$\begin{aligned} 0 &< [\boldsymbol{g}_2' \boldsymbol{g}_2]^2 = [\boldsymbol{g}_2' \boldsymbol{g}]^2 = [\boldsymbol{g}_2' \mathbf{M}_n \boldsymbol{z}_n]^2 \\ &\leqslant (\boldsymbol{g}_2' \mathbf{M}_n \boldsymbol{g}_2)(\boldsymbol{z}_n' \mathbf{M}_n \boldsymbol{z}_n) = \boldsymbol{g}_2' \mathbf{M}_n \boldsymbol{g}_2 \ \mathrm{var}_{\xi_n} g . \end{aligned}$$

Thus $\boldsymbol{g}_2' \mathbf{M}_n \boldsymbol{g}_2 > 0$ and we obtain

$$\mathrm{var}_{\xi_n} g \geqslant \frac{(\boldsymbol{g}_2' \boldsymbol{g}_2)^2}{\boldsymbol{g}_2' \mathbf{M}_n \boldsymbol{g}_2} .$$

It follows that either there is some n_0 that $\mathrm{var}_{\xi_n} g = \infty$ for every $n \geqslant n_0$ or

$$\liminf_{n\to\infty} \mathrm{var}_{\xi_n} g \geqslant (\boldsymbol{g}_2' \boldsymbol{g}_2)^2 / [\lim_{n\to\infty} \boldsymbol{g}_2' \mathbf{M}_n \boldsymbol{g}_2] = \infty .$$

Hence again $\lim_{n\to\infty} \mathrm{var}_{\xi_n} g = \mathrm{var}_\xi g$. □

REMARK. Another proof of Proposition III.14 can be obtained from Propositions IV.4 and IV.22 and from Eq. (15) in Chapter II.

Sometimes a design is changed at one point, say ξ is changed to $(1-\beta)\xi + \beta\xi_x$, where ξ_x is the design concentrated in one point $x \in X$. In such a case

$$\mathbf{M}[(1-\beta)\xi + \beta\xi_x] = (1-\beta)\mathbf{M}(\xi) + \beta \boldsymbol{f}(x)\boldsymbol{f}'(x)$$

and the following proposition can be useful.

PROPOSITION III.15. *Let* $\mathbf{A}$ *be a positive semidefinite matrix; let* $\boldsymbol{f}$ *be a vector and* $\boldsymbol{f} = \boldsymbol{f}_1 + \boldsymbol{f}_2$ *the decomposition of* $\boldsymbol{f}$ *into* $\boldsymbol{f}_1 \in \mathcal{M}(\mathbf{A})$, $\boldsymbol{f}_2 \in \mathcal{N}(\mathbf{A})$. *Let*

$$\mathbf{B} = \mathbf{A} + \boldsymbol{f}\boldsymbol{f}' .$$

Then the matrix $\mathbf{B}$ *is positive semidefinite and*

$$\mathbf{B}^+ = \mathbf{A}^+ - \frac{\mathbf{A}^+\mathbf{f}_1\mathbf{f}_2' + \mathbf{f}_2\mathbf{f}_1'\mathbf{A}^+}{\|\mathbf{f}_2\|^2} + \frac{(1+\mathbf{f}_1'\mathbf{A}^+\mathbf{f}_1)}{\|\mathbf{f}_2\|^4}\mathbf{f}_2\mathbf{f}_2'; \quad (\mathbf{f}\notin\mathcal{M}(\mathbf{A})),$$

$$\mathbf{B}^+ = \mathbf{A}^+ - \frac{\mathbf{A}^+\mathbf{f}_1\mathbf{f}_1'\mathbf{A}^+}{1+\mathbf{f}_1'\mathbf{A}^+\mathbf{f}_1}; \quad (\mathbf{f}\in\mathcal{M}(\mathbf{A})), \tag{13}$$

where $\mathbf{A}^+$, $\mathbf{B}^+$ *are g-inverses of* $\mathbf{A}$, $\mathbf{B}$ *defined in Proposition II.5.*

Proof. If $\mathbf{z}\in\mathcal{M}(\mathbf{B})$, then there is a vector $\mathbf{u}\in R^m$ so that $\mathbf{z} = \mathbf{A}\mathbf{u} + (\mathbf{f}'\mathbf{u})\mathbf{f}$. Using the expressions given in Eq. (13) we can verify that

$$\mathbf{B}\mathbf{B}^+\mathbf{z} = \mathbf{z}. \tag{14}$$

If $\mathbf{z}\in\mathcal{N}(\mathbf{B})$, then $\mathbf{z}\perp\mathcal{M}(\mathbf{A}) = \mathcal{M}(\mathbf{A}^+)$, and $\mathbf{z}\perp\mathbf{f}_2$ since $\mathcal{M}(\mathbf{A})\subset\mathcal{M}(\mathbf{B})$ and $\mathbf{f}\in\mathcal{M}(\mathbf{B})$. Hence, using Eq. (13) we obtain

$$\mathbf{B}^+\mathbf{z} = \mathbf{0}. \tag{15}$$

Comparing Eqs. (14) and (15) with the definition of the g-inverse matrix in Proposition II.5 we finish the proof. □

We shall illustrate the discontinuity of the function $\mathbf{M}(\xi)\in\mathfrak{M}\mapsto \mathrm{var}_\xi g$ in the following example.

Example. Consider the regression model given by

$$X = \langle 0, 2\rangle,$$
$$\Theta = \{\vartheta: \vartheta(x) = \alpha_1 x + \alpha_2 x^2;\ (\alpha_1, \alpha_2)'\in R^2\},$$
$$\sigma^2(x) = 1; \quad (x\in X).$$

Let g be the functional

$$g(\vartheta) \equiv \vartheta(1) = (1, 1)\begin{pmatrix}\alpha_1\\ \alpha_2\end{pmatrix}.$$

Clearly, the design minimizing $\mathrm{var}_\xi g$ is the design ξ_1 concentrated in the point $x = 1$. The corresponding variance is

$$\mathrm{var}_{\xi_1} g = (1, 1)\mathbf{M}^-(\xi_1)\begin{pmatrix}1\\ 1\end{pmatrix} = 1.$$

Now consider another design $\mu(t)$ spread uniformly onto two points $x_1=1+t$, $x_2=1+ct$ for some $t\in\langle -1, 1\rangle$, $c\in\langle -1, 1\rangle$. The corresponding information matrix is

$$\mathbf{M}(\mu(t))=\frac{1}{2}\begin{pmatrix} x_1^2+x_2^2, & x_1^3+x_2^3 \\ x_1^3+x_2^3, & x_1^4+x_2^4 \end{pmatrix}.$$

After simple computation we obtain

$$\det \mathbf{M}(\mu(t))=\frac{1}{4}\, x_1^2x_2^2(x_1-x_2)^2,$$

hence

$$\mathbf{M}^{-1}(\mu(t))=\frac{1}{2\det \mathbf{M}(\mu(t))}\begin{pmatrix} x_1^4+x_2^4, & -x_1^3-x_2^3 \\ -x_1^3-x_2^3, & x_1^2+x_2^2 \end{pmatrix}.$$

Thus

$$\mathrm{var}_{\mu(t)}g=\frac{2}{(1-c)^2}\left[\frac{1}{(1+ct)^2}+\frac{c^2}{(1+t)^2}\right].$$

Now, for a fixed $c\in\langle -1, 1\rangle$ take the limit

$$\lim_{t\to 0} \mathrm{var}_{\mu(t)}g=\frac{2(1+c^2)}{(1-c)^2}.$$

If t tends to 0, then x_1, x_2 tend to the point $x=1$. However, as seen from the last formula, $\mathrm{var}_{\mu(t)}g$ does not tend to $\mathrm{var}_{\xi_1}g$ unless $c=-1$. For other choices of c the limit expression $2(1+c^2)/(1-c)^2$ can achieve any value within the interval $(\mathrm{var}_{\xi_1}g, \infty)$.

EXERCISE III.4. Prove that a sequence of information matrices $\mathbf{M}(\xi_n)$ converges to the information matrix $\mathbf{M}(\xi)$ exactly if for every ϑ, $\vartheta'\in\Theta$

$$\lim_{n\to\infty} \sum_{x\in X} \vartheta(x)\vartheta'(x)\xi_n(x)=\sum_{x\in X} \vartheta(x)\vartheta'(x)\xi(x). \tag{16}$$

EXERCISE III.5. Let g_1, g_2 be two linear functionals defined on the set Θ. Prove that the function $\mathbf{M}(\xi)\mapsto \mathrm{cov}_\xi(g_1, g_2)$ defined on the set

$$\{\mathbf{M}(\xi)\colon \mathbf{g}_1\in\mathcal{M}[\mathbf{M}(\xi)],\ \mathbf{g}_2\in\mathcal{M}[\mathbf{M}(\xi)]\}$$

is continuous at a point $\mathbf{M}(\mu)$ if $\det \mathbf{M}(\mu)\neq 0$.

HINT. Using Proposition II.2 prove that

$$\mathrm{cov}_\xi(g_1, g_2) = \frac{1}{4}\{\mathrm{var}_\xi(g_1+g_2) - \mathrm{var}_\xi(g_1-g_2)\}. \tag{17}$$

III.4. DESIGNS YIELDING SMALL VALUES OF $\mathrm{var}_\xi g$

PROPOSITION III.16 [8]. *Suppose that* $\mathrm{var}_\xi g < \infty$ *and that the vectors belonging to the set* $\{\mathbf{f}(x): \xi(x) > 0\}$ *are linearly dependent. Then there is a design* η *such that* $X_\eta \subsetneqq X_\xi$ *and*

$$\mathrm{var}_\eta g \leqslant \mathrm{var}_\xi g.$$

The design η *can be chosen so that the support* X_η *has at most* m *points.*

PROOF. According to Proposition II.2 there is a vector $\mathbf{z}_\xi \in \mathcal{M}[\mathbf{M}(\xi)]$ such that $\mathbf{M}(\xi)\mathbf{z}_\xi = \mathbf{g}$, $\mathrm{var}_\xi g = \mathbf{z}_\xi'\mathbf{M}(\xi)\mathbf{z}_\xi$. Thus

$$\mathbf{g} = \sum_{x\in X} \mathbf{f}(x)[\mathbf{f}'(x)\mathbf{z}_\xi]\xi(x) \tag{18}$$

and

$$\mathrm{var}_\xi g = \sum_{x\in X} [\mathbf{f}'(x)\mathbf{z}_\xi]^2 \xi(x). \tag{19}$$

Define

$$\xi^*(x) \equiv \xi(x)|\mathbf{f}'(x)\mathbf{z}_\xi|/c,$$

where

$$c \equiv \sum_{x\in X} |\mathbf{f}'(x)\mathbf{z}_\xi|\xi(x). \tag{20}$$

Hence, according to Eq. (18),

$$\mathbf{g} = c\sum_{x\in X} \tilde{\mathbf{f}}(x)\xi^*(x),$$

where we used the notation $\tilde{\mathbf{f}}(x) \equiv \mathbf{f}(x)\,\mathrm{sign}\,[\mathbf{f}'(x)\mathbf{z}_\xi]$. Since the vectors $\tilde{\mathbf{f}}(x)$; $(x \in X_\xi)$ are linearly dependent, there is a function $a: X \mapsto R$ ($a(x) = 0$ for $x \notin X_\xi$) such that

$$\sum_{x\in X} a(x)\tilde{\mathbf{f}}(x) = \mathbf{0}.$$

The sign of $a(.)$ can be chosen so that $A \equiv \sum_{x\in X} a(x) \geqslant 0$. For every

$\lambda \in R$ we have

$$\mathbf{g} = c \sum_{x \in X} \mathbf{f}(x)[\xi^*(x) - \lambda a(x)]. \tag{21}$$

Define

$$\lambda_0 \equiv \min \{\xi^*(x)/a(x) : \xi(x) > 0,\ a(x) > 0\}.$$

It can be easily verified that $\xi^*(x) - \lambda_0 a(x) \geq 0$ for every $x \in X$, and that

$$\sum_{x \in X} [\xi^*(x) - \lambda_0 a(x)] = 1 - \lambda_0 A \leq 1.$$

Define the design

$$\eta(x) \equiv [\xi^*(x) - \lambda_0 a(x)]/(1 - \lambda_0 A); \quad (x \in X).$$

The definition of λ_0 implies $X_\eta \subsetneqq X_\xi$. The expression in Eq. (21) can be written as

$$\mathbf{g} = d \sum_{x \in X} \mathbf{f}(x)\eta(x), \tag{22}$$

where

$$d = c/(1 - \lambda_0 A) \leq c.$$

It follows that the functional g is also estimable under the design η, hence, according to Proposition II.2, there is a vector $\mathbf{z}_\eta \in \mathcal{M}[\mathbf{M}(\eta)]$ such that $\mathbf{g} = \mathbf{M}(\eta)\mathbf{z}_\eta$, $\mathrm{var}_\eta g = \mathbf{z}'_\eta \mathbf{M}(\eta)\mathbf{z}_\eta$. Thus, using the Schwarz inequality

$$\begin{aligned} \mathrm{var}_\eta g &= \sum_{x \in X} |\mathbf{f}'(x)\mathbf{z}_\eta|^2 \eta(x) \\ &\geq \Big[\sum_{x \in X} |\mathbf{f}'(x)\mathbf{z}_\eta|\eta(x)\Big]^2. \end{aligned}$$

On the other hand, it follows from Eq. (22) that

$$\begin{aligned} \mathrm{var}_\eta g = \mathbf{z}'_\eta \mathbf{g} &= d \sum_{x \in X} [\mathbf{f}'(x)\mathbf{z}_\eta]\eta(x) \\ &\leq d \sum_{x \in X} |\mathbf{f}'(x)\mathbf{z}_\eta|\eta(x). \end{aligned}$$

Comparing both bounds for $\mathrm{var}_\eta g$ we obtain

$$\sum_{x\in X} |\mathbf{f}'(x)\mathbf{z}_\eta|\eta(x) \leqslant d \leqslant c = \sum_{x\in X} |\mathbf{f}'(x)\mathbf{z}_\xi|\xi(x)$$

and

$$\mathrm{var}_\eta g \leqslant \left[\sum_{x\in X} |\mathbf{f}'(x)\mathbf{z}_\eta|\eta(x)\right]\left[\sum_{x\in X} |\mathbf{f}'(x)\mathbf{z}_\xi|\xi(x)\right].$$

Thus, according to Eq. (19)

$$\begin{aligned}\mathrm{var}_\xi g &= \sum_{x\in X} |\mathbf{f}'(x)\mathbf{z}_\xi|^2\xi(x)\\ &\geqslant \left[\sum_{x\in X} |\mathbf{f}'(x)\mathbf{z}_\xi|\xi(x)\right]^2\\ &\geqslant \left[\sum_{x\in X} |\mathbf{f}'(x)\mathbf{z}_\eta|\eta(x)\right]\left[\sum_{x\in X} |\mathbf{f}'(x)\mathbf{z}_\xi|\xi(x)\right]\\ &\geqslant \mathrm{var}_\eta g. \;\square\end{aligned}$$

The function $\mathbf{M}(\xi)\in\mathfrak{M}\mapsto \mathrm{var}_\xi g$ is lower semicontinuous on the compact set $\mathfrak{M}$. It follows that this function achieves its minimum, which means that there is at least one design ξ^* with the property

$$\mathrm{var}_{\xi^*} g = \min\{\mathrm{var}_\xi g \colon \xi\in\Xi\}. \tag{23}$$

From Proposition III.16 it follows that the design ξ^* can be chosen so that it is supported by at most m points. From Proposition III.7 it follows that the vectors $\mathbf{f}(x)$; $(x\in X_{\xi^*})$ are on the boundary of the set S defined in Eq. (4). Moreover, we have the following proposition.

PROPOSITION III.17. (*The Elfving theorem* [15].) *The design ξ^* is a solution of Eq. (23) exactly if there is a set $Y\subset X_{\xi^*}$ and a number $c>0$ such that*

a) $c\mathbf{g}$ *is a boundary point of the set S,*

b) $$c\mathbf{g} = \sum_{x\in Y} \mathbf{f}(x)\xi^*(x) - \sum_{x\in X-Y} \mathbf{f}(x)\xi^*(x). \tag{24}$$

In such a case

$$\mathrm{var}_{\xi^*} g = c^{-2} = \inf\{\lambda\colon \lambda>0,\ \lambda^{-1}\mathbf{g}\in S\}.$$

PROOF. 1. Consider a design ξ allowing the estimation of the functional g. According to Proposition II.2 there is $\mathbf{z}_g \in \mathscr{M}[\mathbf{M}(\xi)]$ so that

$$\mathbf{g} = \sum_{x \in X} \mathbf{f}(x)[\mathbf{f}'(x)\mathbf{z}_g]\xi(x).$$

Define the design

$$\xi'(x) \equiv |\mathbf{f}'(x)\mathbf{z}_g|\xi(x) \Big/ \sum_{x \in X} |\mathbf{f}'(x)\mathbf{z}_g|\xi(x); \tag{25}$$

$(x \in X)$.

We can write

$$c_\xi \mathbf{g} = \sum_{x \in X} \mathbf{f}(x) \operatorname{sign}[\mathbf{f}'(x)\mathbf{z}_g]\xi'(x), \tag{26}$$

where

$$c_\xi \equiv \left[\sum_{x \in X} |\mathbf{f}'(x)\mathbf{z}_g|\xi(x)\right]^{-1}.$$

From Eq. (26) it follows, according to Propositions II.2 and II.8, that g is estimable also under the design ξ'; the estimate

$$\sum_{x \in X} c_\xi^{-1} \operatorname{sign}[\mathbf{f}'(x)\mathbf{z}_g]\bar{y}(x)\xi'(x)$$

is an unbiassed estimate of g. (By the symbol $\bar{y}(x)$ we denote the arithmetic mean of the results of the observations in the repeated trial x.) The variance of this estimate is evidently bounded from below by $N \operatorname{var}_{\xi'} g$, hence

$$\begin{aligned} \operatorname{var}_{\xi'} g &\leq N \left\{\sum_{x \in X} c_\xi^{-1} \operatorname{sign}[\mathbf{f}'(x)\mathbf{z}_g]\bar{y}(x)\xi'(x)\right\} = c_\xi^{-2} \\ &= \left(\sum_{x \in X} |\mathbf{f}'(x)\mathbf{z}_g|\xi(x)\right)^2 \\ &\leq \sum_{x \in X} |\mathbf{f}'(x)\mathbf{z}_g|^2\xi(x) = \mathbf{z}_g'\mathbf{M}(\xi)\mathbf{z}_g = \operatorname{var}_\xi g. \end{aligned} \tag{27}$$

We have proven: to every design ξ, such that $\operatorname{var}_\xi g < \infty$, there is a design ξ', such that $\mathrm{var}_\xi g \geq \mathrm{var}_{\xi'} g$, and according to Eq. (26), there is a number $c_\xi > 0$ and a set $Y_\xi \subset X_\xi$ such that

$$c_\xi \mathbf{g} = \sum_{x \in Y_\xi} \mathbf{f}(x)\xi'(x) - \sum_{x \in X - Y_\xi} \mathbf{f}(x)\xi'(x). \tag{28}$$

Denote the set of such "improved" designs by

$$\Xi' \equiv \{\xi' : \xi \in \Xi, \mathrm{var}_\xi g < \infty\}.$$

2. Obviously, if ξ^* is a solution of Eq. (23), then $\xi^* \in \Xi'$. Moreover, in such a case equality signs are achieved in Eq. (27), hence

$$\mathrm{var}_{\xi^*} g = c_{\xi^*}^{-2}.$$

Further, according to Eq. (27)

$$\begin{aligned} c_{\xi^*}^{-2} &\leq \min\{c_\xi^{-2} : \xi \in \Xi, \mathrm{var}_\xi g < \infty\} \\ &\leq \min\{\mathrm{var}_\xi g : \xi \in \Xi\} = c_{\xi^*}^{-2}, \end{aligned}$$

hence

$$c_{\xi^*}^2 = \max\{c_\xi^2 : \xi \in \Xi, \mathrm{var}_\xi g < \infty\}. \tag{29}$$

On the other hand, Eq. (28) implies $c_\xi \mathbf{g} \in S$ whenever $\mathrm{var}_\xi g < \infty$. It follows from Eq. (29) that $c_{\xi^*} g$ is a boundary point of the set S and that

$$c_{\xi^*}^2 = \sup\{\lambda^{-1} : \lambda^{-1} > 0, \lambda^{-1}\mathbf{g} \in S\},$$

that means

$$c_{\xi^*}^{-2} = \inf\{\lambda : \lambda > 0, \lambda^{-1}\mathbf{g} \in S\}.$$

3. Suppose in turn that the design ξ^* has the properties a) and b). Since, according to Eq. (28), every vector belonging to the set $\{c_\xi \mathbf{g} : \xi \in \Xi, \mathrm{var}_\xi g < \infty\}$ is an element of the set S, we obtain from the property a) that

$$\|c\mathbf{g}\|^2 = \max\{\|c_\xi \mathbf{g}\|^2 : \xi \in \Xi, \mathrm{var}_\xi g < \infty\}.$$

Hence, according to Eq. (27)

$$\begin{aligned} c^{-2} &= \min\{c_\xi^{-2} : \xi \in \Xi, \mathrm{var}_\xi g < \infty\} \\ &\leq \min\{\mathrm{var}_\xi g : \xi \in \Xi\}. \end{aligned} \tag{30}$$

On the other hand, the property b) implies that the estimate

$$\sum_{x \in X} c^{-1}[\chi_Y(x) - \chi_{X-Y}(x)]\bar{y}(x)\xi^*(x)$$

is unbiassed for the functional g under the design ξ^*, hence, similarly as in Eq. (27), we obtain

$$\operatorname{var}_{\xi^*} g \leqslant c^{-2} \sum_{x \in X} [\chi_Y(x) - \chi_{X-Y}(x)]^2 \xi^*(x) = c^{-2}.$$

Therefore, from Eq. (30) it follows that ξ^* is a solution of Eq. (23). □

The use of Proposition III.17 for computing optimum experimental designs has been illustrated in Chapter I, Example 2 and Fig. 1.

EXERCISE III.6. Prove: if ξ^* is a solution of Eq. (23), then there is a vector $\mathbf{u} \in R^m$ such that

$$|\mathbf{f}'(x)\mathbf{u}| = \text{const}; \quad (x \in X_{\xi^*}).$$

HINT. The inequality (27) becomes an equality if $\xi = \xi^*$. Take $\mathbf{u} = \mathbf{z}_a$!

Chapter IV

Optimality Criteria in the Regression Model

IV.1. INTRODUCTION

From statistician's point of view good designs ensure small variances of estimates for linear functionals. However, in most cases if we decrease the variance of one functional, the variance of another functional increases. Therefore there is no uniformly best design. It is necessary to formulate the purpose of the experiment more specifically and to choose an adequate optimality criterion.

DEFINITION IV.1. *An optimality criterion is a linear ordering* $\underset{K}{\lesssim}$ *of* Ξ *which has the property*

$$\xi \lesssim \eta \Rightarrow \xi \underset{K}{\lesssim} \eta.$$

Assuming that a function $\phi: \mathfrak{M} \mapsto R \cup \{+\infty\}$ bounded from below is such that

$$\xi \lesssim \eta \Leftrightarrow \phi[\mathbf{M}(\xi)] \leqslant \phi[\mathbf{M}(\eta)], \tag{1}$$

an optimality criterion is defined by

$$\xi \underset{K}{\lesssim} \eta \Leftrightarrow \phi[\mathbf{M}(\xi)] \leqslant \phi[\mathbf{M}(\eta)].$$

The function ϕ is said to be the criterion function, and in turn, the criterion defined by the function ϕ usually is called the ϕ-optimality criterion. A design minimizing $\phi[\mathbf{M}(\xi)]$ is called a ϕ-optimum design.

EXERCISE IV.1. Prove: A function ϕ defining an optimality criterion satisfies

$$\mathbf{M}(\xi) \geq \mathbf{M}(\eta) \Rightarrow \phi[\mathbf{M}(\xi)] \leq \phi[\mathbf{M}(\eta)].$$

HINT. Use Proposition III.2.

Evidently, the same optimality criterion can be expressed by different criteria functions. A careful choice of the function ϕ can be decisive when computing an optimum design. So the function ϕ can have several useful properties (continuity, convexity, differentiality) in some cases. Often the function ϕ can be extended from $\mathfrak{M}$ to a larger set of $m \times m$ matrices, etc.

EXERCISE IV.2. Prove that the function ϕ defined on $\mathfrak{M}$ by

$$\phi[\mathbf{M}(\xi)] = \operatorname{var}_\xi g\,; \quad (\xi \in \Xi)$$

defines an optimality criterion. Show that the same criterion is defined by

$$\phi[\mathbf{M}(\xi)] = \log \operatorname{var}_\xi g\,; \quad (\xi \in \Xi).$$

The optimality criteria are classified into two main groups:

1. the global optimality criteria,
2. the partial optimality criteria.

Global criteria are used when estimating the whole state function $\vartheta \in \Theta$, that is when all parameters $\alpha_1, \ldots, \alpha_m$ are important. On the other hand, partial criteria are used when only a partial information of the state of the observed object is needed, for example, the information about one or some (not all) parameters or some functions of the parameters. It may be that such partial information is obtained from observations performed according to a singular design (i.e. with a singular information matrix). The investigation of analytic properties is more difficult for partial criteria functions. It is also more difficult to check the optimality of a design if it is singular (see Chapter IV.5), or to compute iteratively a partially optimal design (see Chapter V).

Some symbols. We present here some standard symbols used in this and the following chapters.

By $\mathscr{R}^{k \times k}$ we denote the set of all $k \times k$ matrices, and by $\mathfrak{S}^{k \times k}$ the set of all symmetric $k \times k$ matrices.

If $\mathbf{A} \in \mathscr{R}^{k \times k}$, then

$$\operatorname{tr} \mathbf{A} \equiv \sum_{i=1}^{k} \{\mathbf{A}\}_{ii}$$

is the trace of $\mathbf{A}$. By $\langle \mathbf{A}, \mathbf{B} \rangle$ and by $\|\mathbf{A}\|$ we denote the inner product and the norm in $\mathscr{R}^{k \times k}$,

$$\langle \mathbf{A}, \mathbf{B} \rangle \equiv \operatorname{tr} \mathbf{A}'\mathbf{B},$$
$$\|\mathbf{A}\|^2 \equiv \operatorname{tr} \mathbf{A}'\mathbf{A}.$$

Besides, we have the norm of $\mathbf{A}$ considered as an operator in R^k,

$$\|\mathbf{A}\|_\infty \equiv \sup \left\{ \frac{\|\mathbf{A}\boldsymbol{u}\|}{\|\boldsymbol{u}\|} : \boldsymbol{0} \neq \boldsymbol{u} \in R^k \right\}. \tag{2}$$

Observe that the norms $\| \ \|$ and $\| \ \|_\infty$ are topologically equivalent in the finite-dimensional space $\mathscr{R}^{k \times k}$.

When $\mathbf{A}$ is a positive semidefinite matrix and $p > 0$, define the p-th power $\mathbf{A}^p$ and the norm $\|\mathbf{A}\|_p$ as follows. Denote by $\boldsymbol{u}_1, \ldots, \boldsymbol{u}_k$ the orthonormal eigenvectors of the matrix $\mathbf{A}$ and by $a_1, \ldots, a_k$ the corresponding eigenvalues. Set $\mathbf{U} \equiv (\boldsymbol{u}_1, \ldots, \boldsymbol{u}_k)$. We have

$$\mathbf{U}'\mathbf{A}\mathbf{U} = \operatorname{diag}(a_1, \ldots, a_k).$$

Define

$$\mathbf{A}^p \equiv \mathbf{U} \operatorname{diag}(a_1^p, \ldots, a_k^p)\mathbf{U}',$$
$$\|\mathbf{A}\|_p \equiv [k^{-1} \operatorname{tr} \mathbf{A}^p]^{1/p}.$$

If $\mathbf{A} \in \mathfrak{S}^{k \times k}$ and r is a nonnegative integer, then by $\mathbf{A}^{\otimes r}$ we denote the r-th tensor power of $\mathbf{A}$. It is a $k^r \times k^r$ matrix having each row and column labelled by an ordered r-tuplet $i_1, \ldots, i_r$; $(i_l \in \{1, \ldots, k\})$. It is defined by the equality

$$\{\mathbf{A}^{\otimes r}\}_{i_1, \ldots, i_r, j_1, \ldots, j_r} = \prod_{l=1}^{r} \{\mathbf{A}\}_{i_l j_l}.$$

As before, $\mathfrak{M}$ is the set of all information matrices in the experiment. Denote by

$$\mathfrak{M}_+ \equiv \{\mathbf{M} : \mathbf{M} \in \mathfrak{M}, \det \mathbf{M} > 0\}$$

the set of all nonsingular information matrices. Let $\mathscr{L}(\mathfrak{M})$ be the smallest linear subspace of $\mathfrak{S}^{m\times m}$ containing the set $\mathfrak{M}$. If $\mathbf{M}_0 \in \mathfrak{M}$, we denote by $\mathfrak{M}_0$ the set

$$\mathfrak{M}_0 \equiv \left\{\mathbf{M}: \mathbf{M}\in\mathfrak{M}, \frac{1}{2} \det \mathbf{M}_0 < \det \mathbf{M} < 2 \det \mathbf{M}_0\right\}. \tag{3}$$

Evidently, $\mathbf{M}_0 \in \mathfrak{M}_0 \subset \mathfrak{M}_+$.

If ϕ is a function defined on $\mathfrak{M}$, we denote by $\mathfrak{M}_\phi$ the set

$$\mathfrak{M}_\phi \equiv \{\mathbf{M}: \mathbf{M}\in\mathfrak{M}, \phi(\mathbf{M}) < \infty\}.$$

In case of ϕ defined and differentiable in a neighbourhood of $\mathbf{A}$ in the space $\mathscr{R}^{k\times k}$, by $\nabla_{\mathbf{A}}\phi(\mathbf{A})$ (or shortly by $\nabla\phi(\mathbf{A})$) we denote the gradient of the function ϕ at the point $\mathbf{A}$, i.e. the $k\times k$ matrix

$$\{\nabla_{\mathbf{A}}\phi(\mathbf{A})\}_{ij} \equiv \frac{\partial\phi(\mathbf{A})}{\partial\{\mathbf{A}\}_{ij}}; \quad (i, j = 1, \ldots, k).$$

When considering partial optimality criteria it is sometimes necessary to partition an $m\times m$ matrix $\mathbf{A}$ into submatrices

$$\mathbf{A} = \begin{pmatrix} \mathbf{A}_{\mathrm{I}}, & \mathbf{A}_{\mathrm{II}} \\ \mathbf{A}'_{\mathrm{II}}, & \mathbf{A}_{\mathrm{III}} \end{pmatrix}$$

so that $\mathbf{A}_{\mathrm{I}}$ is an $s\times s$ matrix.

EXERCISE IV.3. Let $\mathbf{A}, \mathbf{B} \in \mathscr{R}^{k\times k}$. Prove that

$$\|\mathbf{AB}\| \leq \|\mathbf{A}\| \, \|\mathbf{B}\|.$$

HINT. Use the Schwarz inequality

$$\sum_{l=1}^{k} [\{\mathbf{A}\}_{il}\{\mathbf{B}\}_{lj}]^2 \leq \sum_{l=1}^{k} \{\mathbf{A}\}_{il}^2 \sum_{l=1}^{k} \{\mathbf{B}\}_{lj}^2.$$

IV.2. GLOBAL OPTIMALITY CRITERIA

IV.2.1. *The D-optimality Criterion*

The D-optimality criterion is defined by the criterion function

$$\phi[\mathbf{M}(\xi)] = -\log \det \mathbf{M}(\xi); \quad (\xi \in \Xi).$$

The value of $-\log \det \mathbf{M}(\xi)$ is finite if and only if $\mathbf{M}(\xi)$ is nonsingular, i.e. when all parameters $\alpha_1, \ldots, \alpha_m$ are estimable. (Compare with Proposition II.8!) The ϕ-optimum design (which is called D-optimum) is minimizing the generalized variance of he estimates of $\alpha_1, \ldots, \alpha_m$:

$$\det \mathbf{D}(\hat{\boldsymbol{\alpha}}) = \det \mathbf{M}^{-1}(\xi).$$

EXERCISE IV.4. Let $f_1, \ldots, f_m$ and $h_1, \ldots, h_m$ be two linear bases of the space Θ. Let $\mathbf{M}_f(\xi)$ and $\mathbf{M}_h(\xi)$ be the corresponding information matrices.

Prove that

$$\det \mathbf{M}_f(\xi) \leqslant \det \mathbf{M}_f(\eta) \Leftrightarrow \det \mathbf{M}_h(\xi) \leqslant \det \mathbf{M}_h(\eta).$$

HINT. Find the nonsingular matrix $\mathbf{T}$ such that $\mathbf{T}(f_1, \ldots, f_m)' = (h_1, \ldots, h_m)'$.

As a consequence of Exercise IV.4, by a D-optimal design the generalized variance of the BLUE-s of the parameters is minimized, regardless of the actual parametrization of the regression model.

There is another statistical interpretation of the expression $\det \mathbf{M}^{-1}(\xi)$ when $\hat{\alpha}_1, \ldots, \hat{\alpha}_m$ are distributed according to a multivariate Gaussian probability distribution. First let us prove the following auxiliary statement.

PROPOSITION IV.1. *The volume (the Lebesgue measure) of the m-dimensional ellipsoid*

$$\mathcal{O} \equiv \{\mathbf{z}\colon \mathbf{z} \in R^m, \mathbf{z}'\mathbf{M}(\xi)\mathbf{z} \leqslant c\}$$

is equal to $c^m V(G)[\det \mathbf{M}^{-1}(\xi)]^{1/2}$, *where* $V(G)$ *is the volume of the m-dimensional sphere with a unit radius.*

PROOF. Let $\mathbf{U}$ be the nonsingular matrix such that $\mathbf{U}'\mathbf{M}(\xi)\mathbf{U} = \mathbf{I}$ (Proposition III.12). Set $\mathbf{w} = \mathbf{U}^{-1}\mathbf{z}$ to compute the integral

$$\begin{aligned} V(\mathcal{O}) &= \int_{\mathcal{O}} \mathrm{d}\mathbf{z} = \int_{\{\mathbf{w}\colon \mathbf{w}'\mathbf{w} \leqslant c\}} |\det \mathbf{U}| \, \mathrm{d}\mathbf{w} \\ &= c^m V(G) \, |\det \mathbf{U}|. \end{aligned}$$

From $\mathbf{U}'\mathbf{M}(\xi)\mathbf{U}=\mathbf{I}$ we obtain $(\det \mathbf{U})^2 \det \mathbf{M}(\xi)=1$. Hence

$$V(\mathcal{O})=c^m V(G)\,[\det \mathbf{M}^{-1}(\xi)]^{1/2}. \square$$

Suppose that $(\hat{\alpha}_1, \ldots, \hat{\alpha}_m)'=\hat{\boldsymbol{\alpha}}$ is a Gaussian random vector. Its mean is $\boldsymbol{\alpha}$ and its covariance matrix is $\mathbf{M}^{-1}(\xi)$. The probability density of $\hat{\boldsymbol{\alpha}}$ is

$$f(\hat{\boldsymbol{\alpha}}\,|\,\boldsymbol{\alpha})=\frac{\det^{1/2}\mathbf{M}(\xi)}{(2\pi)^{m/2}}\exp\left\{-\frac{1}{2}[\hat{\boldsymbol{\alpha}}-\boldsymbol{\alpha}]'\mathbf{M}(\xi)[\hat{\boldsymbol{\alpha}}-\boldsymbol{\alpha}]\right\}.$$

Denote by $\mathscr{E}$ the ellipsoid

$$\mathscr{E}\equiv\{\hat{\boldsymbol{\alpha}}:\ \hat{\boldsymbol{\alpha}}\in R^m,\ (\boldsymbol{\alpha}-\hat{\boldsymbol{\alpha}})'\mathbf{M}(\xi)(\boldsymbol{\alpha}-\hat{\boldsymbol{\alpha}})\leqslant c\},$$

where c is a fixed number. Similarly as in Proposition IV.1 we can show that

$$\begin{aligned}&\int_{\mathscr{E}} f(\hat{\boldsymbol{\alpha}}\,|\,\boldsymbol{\alpha})\,\mathrm{d}\hat{\boldsymbol{\alpha}}\\ &=\int_{\{\mathbf{w}:\ \mathbf{w}'\mathbf{w}\leqslant c\}}(2\pi)^{-m/2}\exp\left\{-\frac{1}{2}\mathbf{w}'\mathbf{w}\right\}\mathrm{d}\mathbf{w}\\ &=P\{\chi^2_m\leqslant c\},\end{aligned}$$

where χ^2_m is a χ-square random variable with m degrees of freedom. It follows that the ellipsoid

$$\{\mathbf{z}:\ \mathbf{z}\in R^m,\ (\mathbf{z}-\hat{\boldsymbol{\alpha}})'\mathbf{M}(\xi)(\mathbf{z}-\hat{\boldsymbol{\alpha}})\leqslant c\}$$

covers the true (but unknown) mean $\boldsymbol{\alpha}$ with probability $P\{\chi^2_m\leqslant c\}$. That means it is a confidence ellipsoid. From Proposition IV.1 it follows that a D-optimum design minimizes the volume of the confidence ellipsoid whatever the value of c is.

Due to the good statistical interpretation of D-optimality, D-optimum design were considered in many papers (a survey is in [59]). However, the drawback of the D-optimality criterion is that in some cases the volume of the confidence ellipsoid corresponding to a D-optimum design can be small because it is "narrow but long". That means there is a linear functional which is estimated with a very large variance just under a D-optimum design. Now, let us consider the analytical properties of the D-optimality criterion function.

PROPOSITION IV.2. *The function*

$$\mathbf{M} \in \mathfrak{M} \mapsto \log \det \mathbf{M} \tag{4}$$

is a) *continuous on* $\mathfrak{M}$,

b) *convex on* $\mathfrak{M}$ *and strictly convex on* $\mathfrak{M}_+$,

c) *differentiable whenever it is finite. Its gradient is*

$$\nabla_{\mathbf{M}}[-\log \det \mathbf{M}] = -\mathbf{M}^{-1}. \tag{5}$$

PROOF. a) Directly from the standard definition of a determinant we obtain

$$\det \mathbf{M} = \sum_{\pi} (-1)^{\pi} \{\mathbf{M}\}_{1\pi(1)} \{\mathbf{M}\}_{2\pi(2)} \cdots \{\mathbf{M}\}_{m\pi(m)},$$

where the sum is over all permutations π of the m-tuplet $(1, \ldots, m)$, and where $(-1)^{\pi} = 1$ (or $(-1)^{\pi} = -1$) if the permutation $\pi(1), \ldots, \pi(m)$ is obtained by an even (or an odd) number of interchanges of pairs in the m-tuplet $(1, \ldots, m)$. It follows that $\det \mathbf{M}$ is a homogeneous polynomial of the elements of the matrix $\mathbf{M}$. As a consequence the function $\mathbf{M} \mapsto -\log \det \mathbf{M}$ is continuous.

b) We have to prove that for every $\beta \in \langle 0, 1\rangle$, $\mathbf{M}_1, \mathbf{M}_2 \in \mathfrak{M}$

$$\begin{aligned} &\log \det [(1-\beta)\mathbf{M}_1 + \beta \mathbf{M}_2] \\ &\geqslant (1-\beta) \log \det \mathbf{M}_1 + \beta \log \det \mathbf{M}_2. \end{aligned}$$

This inequality becomes obvious when $\det \mathbf{M}_1 = 0$ or $\det \mathbf{M}_2 = 0$. Hence suppose that $\mathbf{M}_1, \mathbf{M}_2$ are nonsingular. According to Proposition III.12 there is a nonsingular matrix $\mathbf{U}$ such that $\mathbf{U}\mathbf{M}_1\mathbf{U}' = \mathbf{I}$, $\mathbf{U}\mathbf{M}_2\mathbf{U}' = \mathbf{\Lambda} \equiv \operatorname{diag}(\lambda_1, \ldots, \lambda_m)$. Using the strict concavity of the logarithm we obtain

$$\begin{aligned} &\log \det [(1-\beta)\mathbf{M}_1 + \beta \mathbf{M}_2] \\ &= \log \det \mathbf{U}^{-1}[(1-\beta)\mathbf{I} + \beta \mathbf{\Lambda}]\mathbf{U}'^{-1} \\ &\geqslant \log \det \mathbf{U}^{-2} + \sum_{i=1}^{m} [(1-\beta) \log 1 + \beta \log \lambda_i] \\ &= (1-\beta) \log \det \mathbf{U}^{-2} + \beta \log \det \mathbf{U}^{-1}\mathbf{\Lambda}\mathbf{U}'^{-1} \\ &= (1-\beta) \log \det \mathbf{M}_1 + \beta \log \det \mathbf{M}_2. \end{aligned}$$

The inequality is strict if $\beta \in (0, 1)$ and $\lambda_i \neq 1$ at least for one $i \in \{1, \ldots, m\}$, i.e. if $\mathbf{M}_1 \neq \mathbf{M}_2$.

c) When $-\log \det \mathbf{M} < \infty$, $\mathbf{M}$ is nonsingular. The set

$$\{\mathbf{A}: \mathbf{A} \in \mathcal{R}^{m \times m}, |\log \det \mathbf{A} - \log \det \mathbf{M}| < 1\}$$

is a neighbourhood of $\mathbf{M}$ containing only nonsingular matrices because of the continuity of log det. Hence the following derivative is correct:

$$\frac{\mathrm{d}}{\mathrm{d}\{\mathbf{M}\}_{ij}} [\log \det \mathbf{M}] = \frac{1}{\det \mathbf{M}} \frac{\mathrm{d} \det \mathbf{M}}{\mathrm{d}\{\mathbf{M}\}_{ij}}$$

$$= \frac{1}{\det \mathbf{M}} \frac{\mathrm{d}}{\mathrm{d}\{\mathbf{M}\}_{ij}} \left[\sum_{k=1}^{m} \{\mathbf{M}\}_{ik} \det \mathbf{M}^{(ik)} \right] = \frac{\det \mathbf{M}^{(ij)}}{\det \mathbf{M}} = \{\mathbf{M}^{-1}\}_{ji},$$

where $\det \mathbf{M}^{(ik)}$ is the algebraic complement of $\{\mathbf{M}\}_{ik}$ in the matrix $\mathbf{M}$. □

IV.2.2. *The A-optimality Criterion*

The criterion of A-optimality is defined by the criterion function

$$\phi[\mathbf{M}(\xi)] = \sum_{i=1}^{m} \mathrm{var}_\xi \alpha_i\,; \quad (\xi \in \Xi). \tag{6}$$

Hence an A-optimum design suppresses the variances of the estimates $\hat{\alpha}_1, \ldots, \hat{\alpha}_m$ but it does not take the correlation between these estimates into account.

EXERCISE IV.5. Show that instead of (6) we can write

$$\begin{aligned} \phi[\mathbf{M}(\xi)] &= \mathrm{tr}\, \mathbf{M}^{-1}(\xi) \quad \text{if } \det \mathbf{M}(\xi) \neq 0, \\ &= \infty \qquad\qquad\quad \text{if } \det \mathbf{M}(\xi) = 0. \end{aligned}$$

HINT. Use Proposition II.8 and Eq. (24) in Chapter II.

PROPOSITION IV.3. *The function*

$$\mathbf{M}(\xi) \in \mathfrak{M} \mapsto \sum_{i=1}^{m} \mathrm{var}_\xi \alpha_i \quad \textit{is}$$

a) continuous on $\mathfrak{M}$,

b) convex on $\mathfrak{M}$ *and strictly convex on* $\mathfrak{M}_+$,

c) *differentiable whenever it is finite, and*

$$\nabla \operatorname{tr} \mathbf{M}^{-1}(\xi) = -\mathbf{M}^{-2}(\xi). \tag{7}$$

PROOF

a) If $\sum_{i=1}^{m} \operatorname{var}_\xi \alpha_i < \infty$, then $\operatorname{var}_\xi \alpha_i < \infty$ for $i = 1, \ldots, m$, and the continuity at the point $\mathbf{M}(\xi)$ follows from Proposition III.14.

In case of $\sum_{i=1}^{m} \operatorname{var}_\xi \alpha_i = \infty$ consider a sequence $\{\mathbf{M}(\xi_n)\}_{n=1}^{\infty}$ tending to $\mathbf{M}(\xi)$. There is at least one $i \in \{1, \ldots, m\}$ such that $\operatorname{var}_\xi \alpha_i = \infty$ hence, according to Proposition III.14

$$\lim_{n\to\infty} \operatorname{var}_{\xi_n} \alpha_i = \infty.$$

It follows that

$$\liminf_{n\to\infty} \sum_{i=1}^{m} \operatorname{var}_{\xi_n} \alpha_i \geq \lim_{n\to\infty} \operatorname{var}_{\xi_n} \alpha_i = \infty.$$

b) The convexity is a direct consequence of Proposition III.13.

c) If $\sum_{i=1}^{m} \operatorname{var}_\xi \alpha_i < \infty$, then the parameters $\alpha_1, \ldots, \alpha_m$ are estimable, hence the matrix $\mathbf{M}(\xi)$ is nonsingular, and

$$\sum_{i=1}^{m} \operatorname{var}_\xi \alpha_i = \operatorname{tr} \mathbf{M}^{-1}(\xi).$$

Let us take the derivatives of

$$\mathbf{I} = \mathbf{M}\mathbf{M}^{-1}$$

with respect to the elements of the matrix $\mathbf{M}$. We obtain

$$\mathbf{0} = \frac{\mathbf{M}}{\partial\{\mathbf{M}\}_{ij}} \mathbf{M}^{-1} + \mathbf{M} \frac{\partial \mathbf{M}^{-1}}{\partial\{\mathbf{M}\}_{ij}};$$
$$(i, j = 1, \ldots, m)$$

and

$$\frac{\partial \mathbf{M}^{-1}}{\partial\{\mathbf{M}\}_{ij}} = -\mathbf{M}^{-1} \frac{\partial \mathbf{M}}{\partial\{\mathbf{M}\}_{ij}} \mathbf{M}^{-1};$$
$$(i, j = 1, \ldots, m).$$

Hence

$$\nabla_{\mathbf{M}} \operatorname{tr} \mathbf{M}^{-1} = -\mathbf{M}^{-2}. \ \square$$

EXERCISE IV.6. Let the functions $f_1, \ldots, f_m$ be orthonormal with respect to a measure w, i.e.

$$\int_X f_i(x) f_j(x) w(\mathrm{d}x) = 1 \ \text{if} \ i = j,$$
$$= 0 \ \text{if} \ i \neq j.$$

Prove that the design ξ^* is optimum according to the A-optimality criterion exactly if

$$\int_X (\operatorname{var}_{\xi^*} g_x) w(\mathrm{d}x) = \min_{\xi} \int_X (\operatorname{var}_{\xi} g_x) w(\mathrm{d}x),$$

where $g_x(\vartheta) = \vartheta(x)$; $(x \in X, \ \vartheta \in \Theta)$.

HINT. Verify that

$$\int_X (\operatorname{var}_{\xi} g_x) w(\mathrm{d}x) = \operatorname{tr} \mathbf{M}^{-1}(\xi).$$

IV.2.3. *The G-optimality Criterion*

The G-optimality criterion is defined by the criterion function

$$\phi \colon \mathbf{M}(\xi) \mathfrak{M} \mapsto \sup_{x \in X} \operatorname{var}_{\xi} g_x, \tag{8}$$

where for every $x \in X$, g_x is the functional

$$g_x(\vartheta) = \vartheta(x); \quad (\vartheta \in \Theta).$$

Evidently,

$$\begin{aligned} \phi[\mathbf{M}(\xi)] &= \max_{x \in X} \boldsymbol{f}'(x) \mathbf{M}^{-1}(\xi) \boldsymbol{f}(x) \quad && \text{if } \det \mathbf{M}(\xi) \neq 0, \\ &= \infty && \text{if } \det \mathbf{M}(\xi) = 0. \end{aligned} \tag{9}$$

The experimenter optimizing the design according to the G-optimality criterion intends to get a good estimate of the whole state-function $\vartheta \in \Theta$.

Consider now the following auxiliary proposition.

PROPOSITION IV.4. *Suppose that* $\mathbf{M}_0 \in \mathfrak{M}$, $\det \mathbf{M}_0 \neq 0$. *Let* $C \subset R^m$ *be a compact set. Take* $\mathfrak{M}_0$ *as in Eq. (3). Then the matrix-valued function*

$$\mathbf{M} \in \mathfrak{M}_0 \mapsto \mathbf{M}^{-1}$$

and the real-valued function

$$\mathbf{M} \in \mathfrak{M}_0 \mapsto \max_{c \in C} \mathbf{c}'\mathbf{M}^{-1}\mathbf{c}$$

are continuous at the point $\mathbf{M}_0$.

PROOF. Take $\mathbf{M}_1 \in \mathfrak{M}_0$, $\|\mathbf{M}_1 - \mathbf{M}_0\| < \delta$. We have

$$\|\mathbf{M}_1\| \leq \|\mathbf{M}_1 - \mathbf{M}_0\| + \|\mathbf{M}_0\| < \delta + \|\mathbf{M}_0\|.$$

Denote the orthonormal eigenvectors and the eigenvalues of $\mathbf{M}_1$ by $\mathbf{v}_1, \ldots, \mathbf{v}_m$ and by $\lambda_1, \ldots, \lambda_m$. Denote $\mathbf{V} \equiv (\mathbf{v}_1, \ldots, \mathbf{v}_m)$. Then $\mathbf{V}'\mathbf{M}_1\mathbf{V} = \operatorname{diag}(\lambda_1, \ldots, \lambda_m)$, and (see Exercise IV.3)

$$\sum_{i=1}^{m} \lambda_i^2 \leq \|\mathbf{V}\|^4 \|\mathbf{M}_1\|^2 \leq m^2(\delta + \|\mathbf{M}_0\|)^2$$

since $\|\mathbf{V}\|^2 = \operatorname{tr} \mathbf{V}'\mathbf{V} = \operatorname{tr} \mathbf{I} = m$. Therefore

$$\lambda_i \leq m(\delta + \|\mathbf{M}_0\|); \quad (i = 1, \ldots, m).$$

On the other hand

$$\det \mathbf{M}_1 = \prod_{i=1}^{m} \lambda_i.$$

Hence, using Eq. (3), we obtain that the minimum eigenvalue of $\mathbf{M}_1$, say λ_{i_0}, is equal to

$$\lambda_{i_0} = \frac{\det \mathbf{M}_1}{\prod_{i \neq i_0} \lambda_i} > \frac{(\det \mathbf{M}_0)/2}{[m(\delta + \|\mathbf{M}_0\|)]^{m-1}} \equiv d > 0$$

and

$$\|\mathbf{M}_1^{-1}\|^2 = \operatorname{tr}(\mathbf{V}\mathbf{\Lambda}^{-1}\mathbf{V}'\mathbf{V}\mathbf{\Lambda}^{-1}\mathbf{V}') \leq \|\mathbf{V}\mathbf{\Lambda}^{-1}\mathbf{V}'\|^2 \leq m^2\|\mathbf{\Lambda}^{-1}\|^2$$
$$< m^3\lambda_{i_0}^{-2} < m^3 d^{-2}.$$

Thus

$$\|\mathbf{M}_1^{-1} - \mathbf{M}_0^{-1}\| = \|\mathbf{M}_1^{-1}(\mathbf{M}_1 - \mathbf{M}_0)\mathbf{M}_0^{-1}\| < \|\mathbf{M}_0^{-1}\| m^{3/2} d^{-1} \delta. \tag{10}$$

This implies the required continuity of the mapping $\mathbf{M} \in \mathfrak{M}_0 \mapsto \mathbf{M}^{-1}$ at the point $\mathbf{M}_0$.

Define the function

$$\varphi: (\mathbf{M}, \boldsymbol{c}) \in \mathfrak{M}_0 \times C \mapsto \boldsymbol{c}'\mathbf{M}^{-1}\boldsymbol{c}.$$

The function φ is the product of continuous functions, hence it is continuous on $\mathfrak{M}_0 \times C$. Take $\mathbf{M}_1 \in \mathfrak{M}_0$. Define the points $\boldsymbol{c}_0 \in C$ and $\boldsymbol{c}_1 \in C$ by the equations

$$\varphi(\mathbf{M}_0, \boldsymbol{c}_0) = \max_{c \in C} \varphi(\mathbf{M}_0, \boldsymbol{c}),$$

$$\varphi(\mathbf{M}_1, \boldsymbol{c}_1) = \max_{c \in C} \varphi(\mathbf{M}_1, \boldsymbol{c}).$$

Denote $K \equiv \max\{\|\boldsymbol{c}\| : \boldsymbol{c} \in C\}$. Take $\varepsilon > 0$ and denote $\eta \equiv \varepsilon / K^2$. From Eq. (10) it follows that there is a number $\delta > 0$ such that $\|\mathbf{M}_1 - \mathbf{M}_0\| < \delta$ implies $\|\mathbf{M}_1^{-1} - \mathbf{M}_0^{-1}\| < \eta$. Hence

$$|\varphi(\mathbf{M}_1, \boldsymbol{c}) - \varphi(\mathbf{M}_0, \boldsymbol{c})| \leqslant \|\boldsymbol{c}\|^2 \|\mathbf{M}_1^{-1} - \mathbf{M}_0^{-1}\| \leqslant K^2 \eta < \varepsilon$$

for every $\boldsymbol{c} \in C$. It follows that

$$\varphi(\mathbf{M}_0, \boldsymbol{c}_0) < \varphi(\mathbf{M}_1, \boldsymbol{c}_0) + \varepsilon < \varphi(\mathbf{M}_1, \boldsymbol{c}_1) + \varepsilon,$$

$$\varphi(\mathbf{M}_1, \boldsymbol{c}_1) < \varphi(\mathbf{M}_0, \boldsymbol{c}_1) + \varepsilon < \varphi(\mathbf{M}_0, \boldsymbol{c}_0) + \varepsilon.$$

Thus

$$|\varphi(\mathbf{M}_0, \boldsymbol{c}_0) - \varphi(\mathbf{M}_1, \boldsymbol{c}_1)| < \varepsilon. \;\square$$

PROPOSITION IV.5. *The function ϕ defined by Eq. (9) is*

a) *continuous on $\mathfrak{M}$,*

b) *convex on $\mathfrak{M}$ and strictly convex on $\mathfrak{M}_+$.*

PROOF. a) If $\sup_{x \in X} \operatorname{var}_\xi g_x = \infty$, then there is $x_0 \in X$ so that $\operatorname{var}_\xi g_{x_0} = \infty$.

Indeed, assuming that $\operatorname{var}_\xi g_x < \infty$ for every $x \in X$, we obtain that every functional g_x is estimable. It follows that $\boldsymbol{f}(x) \in \mathcal{M}[\mathbf{M}(\xi)]$; $(x \in X)$

(see Proposition II.7), therefore $\mathcal{M}[\mathbf{M}(\xi)] = R^m$, and $\mathbf{M}(\xi)$ is nonsingular. As a consequence, the function

$$x \in X \mapsto \boldsymbol{f}'(x)\mathbf{M}^{-1}(\xi)\boldsymbol{f}(x) = \mathrm{var}_\xi g_x$$

is continuous, hence bounded on the compact set X.

Suppose now that $\mathrm{var}_\xi g_{x_0} = \infty$ and that

$$\lim_{n\to\infty} \mathbf{M}(\xi_n) = \mathbf{M}(\xi).$$

From Proposition III.14 it follows that

$$\lim_{n\to\infty} \mathrm{var}_{\xi_n} g_{x_0} = \infty,$$

hence

$$\lim_{n\to\infty} \sup_{x\in X} \mathrm{var}_{\xi_n} g_x = \infty = \sup_{x \in X} \mathrm{var}_\xi g_x .$$

When $\det \mathbf{M}_0 \neq 0$, the set $\mathfrak{M}_0$ defined by Eq. (3) is an open subset of $\mathfrak{M}_+$. Therefore

$$\phi(\mathbf{M}) = \max_{x \in X} \boldsymbol{f}'(x)\mathbf{M}^{-1}\boldsymbol{f}(x); \quad (\mathbf{M} \in \mathfrak{M}_0),$$

hence the continuity of ϕ on $\mathfrak{M}_+$ is a consequence of Proposition IV.4.

b) The convexity of ϕ follows from Proposition III.13. Denote by x_0 the point of maximum of the function

$$x \in X \mapsto \boldsymbol{f}'(x)[(1-\beta)\mathbf{M}_1 + \beta\mathbf{M}_2]^{-1}\boldsymbol{f}(x).$$

From Proposition III.13 it follows that

$$\begin{aligned}
&\boldsymbol{f}'(x_0)[(1-\beta)\mathbf{M}_1 + \beta\mathbf{M}_2]^{-1}\boldsymbol{f}(x_0) \\
&< (1-\beta)\boldsymbol{f}'(x_0)\mathbf{M}_1^{-1}\boldsymbol{f}(x_0) + \beta\boldsymbol{f}'(x_0)\mathbf{M}_2^{-1}\boldsymbol{f}(x_0) \\
&\leq (1-\beta)\max_{x\in X} \boldsymbol{f}'(x)\mathbf{M}_1^{-1}\boldsymbol{f}(x) + \beta \max_{x \in X} \boldsymbol{f}'(x)\mathbf{M}_2^{-1}\boldsymbol{f}(x),
\end{aligned}$$

i.e. the strict convexity of ϕ. □

IV.2.4. *The Equivalence of D-optimum and G-optimum Designs*

The D-optimality and the G-optimality criteria are different criteria with clearly different statistical interpretations. Therefore the theorem stating that both criteria are giving the same optimum designs (cf. [2]) attracted the attention of statisticians. We shall prove this theorem.

PROPOSITION IV.6. *Assuming that* $\sigma^2(x) = const$; $(x \in X)$, *a design* μ *is D-optimal if and only if it is G-optimal, i.e. a) and b) are equivalent statements*:

a) $$\det \mathbf{M}(\mu) = \max \{\det \mathbf{M}(\xi) : \xi \in \Xi\},$$

b) $$\max_{x \in X} \mathbf{f}'(x)\mathbf{M}^{-1}(\mu)\mathbf{f}(x)$$
$$= \min \{\max_{x \in X} \mathbf{f}'(x)\mathbf{M}^{-1}(\xi)\mathbf{f}(x) : \xi \in \Xi, \det \mathbf{M}(\xi) \neq 0\}.$$

Moreover, the last expression is equal to m (= *the number of parameters*).

PROOF. 1. Suppose first that μ is D-optimal, i.e. that the function $M \in \mathfrak{M} \mapsto -\log \det \mathbf{M}$ achieves its minimum at $\mathbf{M}(\mu)$. According to Proposition IV.2, this function is finite, continuous and differentiable on the set

$$\{\mathbf{A} : \mathbf{A} \in \mathscr{R}^{k \times k}, |\log \det \mathbf{A} - \log \det \mathbf{M}(\mu)| < 1\}$$

and

$$\nabla\{-\log \det \mathbf{M}(\mu)\} = -\mathbf{M}^{-1}(\mu).$$

It follows that the function

$$\beta \in (0, 1\rangle \mapsto -\log \det [(1-\beta)\mathbf{M}(\mu) + \beta \mathbf{f}(x)\mathbf{f}'(x)]$$

is finite, continuous and differentiable, and it achieves its minimum at the point $\beta = 0$. Using Eq. (5) we obtain

$$0 \leq \left. \frac{\partial\{-\log \det [(1-\beta)\mathbf{M}(\mu) + \beta \mathbf{f}(x)\mathbf{f}'(x)]\}}{\partial \beta} \right|_{\beta=0}$$

$$= -\operatorname{tr}\,[\nabla \log \det \mathbf{M}(\mu)][\mathbf{f}(x)\mathbf{f}'(x) - \mathbf{M}(\mu)]$$
$$= -\mathbf{f}'(x)\mathbf{M}^{-1}(\mu)\mathbf{f}(x) + m,$$

that is

$$\max_{x\in X} \mathbf{f}'(x)\mathbf{M}^{-1}(\mu)\mathbf{f}(x) \leq m. \tag{11}$$

On the other hand, we have

$$\max_{x\in X} \mathbf{f}'(x)\mathbf{M}^{-1}(\xi)\mathbf{f}(x) \geq \sum_{x\in X} \mathbf{f}'(x)\mathbf{M}^{-1}(\xi)\mathbf{f}(x)\xi(x)$$
$$= \operatorname{tr}\,\mathbf{M}^{-1}(\xi)\mathbf{M}(\xi) = m,$$

for every nonsingular design ξ. Hence μ is G-optimal, and

$$\max_{x\in X} \mathbf{f}'(x)\mathbf{M}^{-1}(\mu)\mathbf{f}(x) = m. \tag{12}$$

2. Suppose now that μ is G-optimal.
Using Proposition II.7 we obtain that $\mathbf{M}(\mu)$ is a nonsingular matrix. Let ξ^* be any D-optimum design. Then

$$1 \leq \det \mathbf{M}(\xi^*)\mathbf{M}^{-1}(\mu) = \prod_{i=1}^{m} \lambda_i, \tag{13}$$

where $\lambda_1, \ldots, \lambda_m$ are the eigenvalues of the matrix $\mathbf{M}(\xi)\mathbf{M}^{-1}(\mu)$. Using the well-known inequality between the arithmetic mean and the geometric mean we obtain from Eq. (12)

$$\left(\prod_{i=1}^{m} \lambda_i\right)^{1/m} \leq \frac{1}{m}\sum_{i=1}^{m} \lambda_i = \frac{1}{m}\operatorname{tr}\,\mathbf{M}(\xi^*)\mathbf{M}^{-1}(\mu)$$
$$= \frac{1}{m}\sum_{x\in X} \mathbf{f}'(x)\mathbf{M}^{-1}(\mu)\mathbf{f}(x)\xi^*(x)$$
$$\leq \frac{1}{m}\max_{x\in X} \mathbf{f}'(x)\mathbf{M}^{-1}(\mu)\mathbf{f}(x) = 1. \tag{14}$$

Comparing Eqs. (13) and (14) we obtain

$$\det \mathbf{M}(\mu) = \det \mathbf{M}(\xi^*),$$

hence μ is a D-optimum design. □

NOTES. D-optimum and G-optimum designs are no more equivalent if the function $x \in X \mapsto \sigma^2(x)$ is not constant. According to the note in the introduction to Chapter III, Proposition IV.6 implies that μ is D-optimum exactly if

$$\max_{x \in X} \sigma^{-2}(x)\mathbf{f}'(x)\mathbf{M}^{-1}(\mu)\mathbf{f}(x)$$
$$= \min_{\xi \in \Xi} \max_{x \in X} \sigma^{-2}(x)\mathbf{f}'(x)\mathbf{M}^{-1}(\xi)\mathbf{f}(x). \tag{15}$$

Another proof of Eq. (15) and of Proposition IV.6 follows from Proposition IV.27.

IV.2.5. *The E-optimality Criterion*

The E-optimality criterion is defined by the criterion function

$$\phi[\mathbf{M}(\xi)] = \sup \{\text{var}_\xi g : \|\mathbf{g}\| = 1\}; \quad (\xi \in \Xi). \tag{16}$$

Denote the minimum eigenvalue of $\mathbf{M}(\xi)$ by λ_ξ. Let

$$G_1 \equiv \{\mathbf{g} : \mathbf{g} \in R^m, \|\mathbf{g}\| = 1\}.$$

PROPOSITION IV.7. *The function ϕ defined by Eq. (16) can be expressed as*

$$\begin{aligned} \phi[\mathbf{M}(\xi)] &= \lambda_\xi^{-1} \text{ if } \det \mathbf{M}(\xi) \neq 0, \\ &= \infty \quad \text{if } \det \mathbf{M}(\xi) = 0. \end{aligned} \tag{17}$$

PROOF. According to Eq. (16), $\phi[\mathbf{M}(\xi)] = \infty$ exactly if there is at least one functional which is not estimable under ξ, that is exactly if $\mathbf{M}(\xi)$ is singular (Proposition II.7).

In case of $\det \mathbf{M}(\xi) \neq 0$ denote the eigenvalues and the orthonormal eigenvectors of $\mathbf{M}(\xi)$ by $\lambda_1, \ldots, \lambda_m$ and by $\mathbf{u}_1, \ldots, \mathbf{u}_m$. We have

$$\mathbf{M}^{-1}(\xi)\mathbf{u}_i = \mathbf{M}^{-1}(\xi)\mathbf{M}(\xi)\mathbf{u}_i\lambda_i^{-1} = \lambda_i^{-1}\mathbf{u}_i;$$
$$(i = 1, \ldots, m).$$

Hence

$$\sup\{\mathbf{g}'\mathbf{M}^{-1}(\xi)\mathbf{g}\colon \mathbf{g}\in G_1\}\geq \mathbf{u}_i'\mathbf{M}^{-1}(\xi)\mathbf{u}_i=\lambda_i^{-1};\quad (i=1,\ldots,m),$$

thus

$$\lambda_\xi^{-1}\leq \sup\{\mathrm{var}_\xi g\colon \mathbf{g}\in G_1\}. \tag{18}$$

Every vector $\mathbf{g}\in G_1$ can be decomposed as

$$\mathbf{g}=\sum_{i=1}^m c_i\mathbf{u}_i,$$

where $\sum_{i=1}^m c_i^2=1$. Hence

$$\mathbf{g}'\mathbf{M}^{-1}(\xi)\mathbf{g}=\sum_{i,j=1}^m c_i\mathbf{u}_i'\mathbf{M}^{-1}(\xi)\mathbf{u}_jc_j\leq\lambda_\xi^{-1}.$$

Comparing with (18) we obtain

$$\sup\{\mathrm{var}_\xi g\colon \mathbf{g}\in G_1\}=\lambda_\xi^{-1}.\ \square$$

PROPOSITION IV.8. *The function ϕ defined by Eq. (16) is*
a) *continuous on $\mathfrak{M}$,*
b) *convex on $\mathfrak{M}$ and strictly convex on $\mathfrak{M}_+$.*

PROOF. a) Suppose that

$$\lim_{n\to\infty}\mathbf{M}(\xi_n)=\mathbf{M}(\xi).$$

If $\phi[\mathbf{M}(\xi)]=\infty$, then there is a vector $\mathbf{g}\notin\mathcal{M}[\mathbf{M}(\xi)]$; hence, according to Proposition III.14, $\lim_{n\to\infty}\mathrm{var}_{\xi_n}g=\mathrm{var}_\xi g=\infty$. From

$$\phi[\mathbf{M}(\xi_n)]\geq\mathrm{var}_{\xi_n}g;\quad (n=1,2,\ldots)$$

it follows that

$$\lim_{n\to\infty}\phi[\mathbf{M}(\xi_n)]=\infty.$$

If $\det\mathbf{M}(\xi)\neq 0$ we can write

$$\phi(\mathbf{M}) = \max \{\mathbf{g}'\mathbf{M}^{-1}\mathbf{g} : \mathbf{g} \in G_1\}$$

for every $\mathbf{M}$ belonging to the set

$$\mathfrak{M}_0 = \{\mathbf{M} : \mathbf{M} \in \mathfrak{M}, \tfrac{1}{2} \det \mathbf{M}(\xi) < \det \mathbf{M} < 2 \det \mathbf{M}(\xi)\}.$$

(Observe that G_1 is compact, hence sup = max.) The continuity of ϕ on $\mathfrak{M}_0$ follows from Proposition IV.4. Since $\mathfrak{M}_0$ is open in $\mathfrak{M}$, the function ϕ is continuous on $\mathfrak{M}$ at the point $\mathbf{M}$.

b) Let $\mathbf{M}_1, \mathbf{M}_2 \in \mathfrak{M}$, $\det \mathbf{M}_1 \neq 0$, $\det \mathbf{M}_2 \neq 0$, $\beta \in (0, 1)$. Denote by $\mathbf{g}_0$ the vector corresponding to the maximum of the function

$$\mathbf{g} \in G_1 \mapsto \mathbf{g}'[(1-\beta)\mathbf{M}_1 + \beta\mathbf{M}_2]^{-1}\mathbf{g}.$$

It follows from Proposition III.13 that

$$\begin{aligned} \phi[(1-\beta)\mathbf{M}_1 + \beta\mathbf{M}_2] &= \mathbf{g}_0'[(1-\beta)\mathbf{M}_1 + \beta\mathbf{M}_2]^{-1}\mathbf{g}_0 \\ &< (1-\beta)\mathbf{g}_0'\mathbf{M}_1^{-1}\mathbf{g}_0 + \beta\mathbf{g}_0\mathbf{M}_2^{-1}\mathbf{g}_0 \\ &\leqslant (1-\beta)\phi(\mathbf{M}_1) + \beta\phi(\mathbf{M}_2), \end{aligned}$$

which implies the strict convexity of ϕ on the set $\mathfrak{M}_+$. The convexity on the set $\mathfrak{M}$ follows, since $\det \mathbf{M} = 0$ implies $\phi(\mathbf{M}) = \infty$. □

IV.2.6. *Linear Optimality Criteria* [13]

These criteria are defined by criteria functions of the form

$$\begin{aligned} \phi(\mathbf{M}) &= \operatorname{tr} \mathbf{W}\mathbf{M}^{-1} \quad \text{if } \det \mathbf{M} \neq 0, \\ &= \infty \qquad\qquad\; \text{if } \det \mathbf{M} = 0, \end{aligned} \tag{19}$$

where $\mathbf{W}$ is a positive definite $m \times m$ matrix. The A-optimality criterion corresponds to the particular case of $\mathbf{W} = \mathbf{I}$. If $\mathbf{W}$ is a diagonal matrix, then

$$\phi[\mathbf{M}(\xi)] = \sum_{i=1}^{m} \{\mathbf{W}\}_{ii} \operatorname{var}_\xi \alpha_i ; \quad (\xi \in \Xi),$$

i.e. ϕ is the weighted sum of the variances.

The optimality criterion given by Eq. (19) is said to be linear because of the linearity of ϕ on the set $\{\mathbf{M}^{-1} : \mathbf{M} \in \mathfrak{M}_+\}$.

PROPOSITION IV.9. *Let ϕ^* be a linear function defined on $\mathfrak{S}^{m\times m}$ that is nonnegative (or positive) on the set of positive semidefinite (or positive definite) matrices. Then there is a positive semidefinite (or definite) matrix $\mathbf{W}\in\mathfrak{S}^{m\times m}$ such that*

$$\phi^*(\mathbf{A})=\operatorname{tr}\mathbf{WA};\quad (\mathbf{A}\in\mathfrak{S}^{m\times m}). \tag{20}$$

PROOF. The linear space $\mathfrak{S}^{m\times m}$ is a Hilbert space with the inner product

$$\langle\mathbf{A},\mathbf{B}\rangle=\operatorname{tr}\mathbf{AB};\quad (\mathbf{A},\mathbf{B}\in\mathfrak{S}^{m\times m}).$$

Therefore, according to the Riesz representation theorem (cf. [100] or Chapter VII.1) there is a unique matrix $\mathbf{W}\in\mathfrak{S}^{m\times m}$ satisfying Eq. (20). Take $\boldsymbol{v}\in R^m$, $\boldsymbol{v}\neq\boldsymbol{0}$. The matrix $\boldsymbol{vv}'$ is positive semidefinite, hence

$$\boldsymbol{v}'\mathbf{W}\boldsymbol{v}=\phi^*(\boldsymbol{vv}')\geq 0\quad (\text{or } >0)$$

and so $\mathbf{W}$ is positive semidefinite (or definite). □

PROPOSITION IV.10. *The function ϕ defined by Eq. (19) is*
 a) *continuous on $\mathfrak{M}$,*
 b) *convex on $\mathfrak{M}$ and strictly convex on $\mathfrak{M}_+$,*
 c) *differentiable whenever it is finite, and*

$$\nabla_{\mathbf{M}}\operatorname{tr}\mathbf{WM}^{-1}=-\mathbf{M}^{-1}\mathbf{WM}^{-1}. \tag{21}$$

PROOF. a) According to Proposition III.12 there is a nonsingular matrix $\mathbf{U}$ such that

$$\mathbf{U}'\mathbf{WU}=\mathbf{I}.$$

We can write

$$\begin{aligned}\phi(\mathbf{M})&=\operatorname{tr}(\mathbf{U}'\mathbf{MU})^{-1} && \text{if } \det(\mathbf{U}'\mathbf{MU})\neq 0,\\ &=\infty && \text{otherwise.}\end{aligned}$$

The mapping

$$\Psi_1:\mathbf{M}\in\mathfrak{M}\mapsto\mathbf{U}'\mathbf{MU}$$

is linear and it is continuous. The set $\Psi_1(\mathfrak{M})=\{\mathbf{U}'\mathbf{MU}:\mathbf{M}\in\mathfrak{M}\}$ is the set of all information matrices in the considered experiment $(X,\Theta,1)$

but with a changed linear basis of Θ: $(\mathbf{U}'\mathbf{f})_1, \ldots, (\mathbf{U}'\mathbf{f})_m$. This can be seen from

$$\mathbf{U}'\mathbf{M}(\xi)\mathbf{U} = \sum_{x \in X} [\mathbf{U}'\mathbf{f}(x)][\mathbf{U}'\mathbf{f}(x)]'\xi(x).$$

Define another function Ψ_2 on the set $\Psi_1(\mathfrak{M})$ by

$$\begin{aligned} \Psi_2(\mathbf{U}'\mathbf{M}\mathbf{U}) &= \operatorname{tr}(\mathbf{U}'\mathbf{M}\mathbf{U})^{-1} \quad \text{if } \det(\mathbf{U}'\mathbf{M}\mathbf{U}) \neq 0, \\ &= \infty \qquad\qquad\qquad\quad \text{if } \det(\mathbf{U}'\mathbf{M}\mathbf{U}) = 0. \end{aligned}$$

According to Proposition IV. 3, the function Ψ_2 is continuous, which implies the continuity of $\phi = \Psi_2 \circ \Psi_1$.

b) The required convexity of ϕ follows from the convexity of Ψ_2 (see Proposition IV.3) and from the strict convexity of Ψ_2 on the set $\{\Psi_1(\mathbf{M}) : \mathbf{M} \in \mathfrak{M}, \det \mathbf{M} \neq 0\}$.

c) Using Proposition IV.3 we obtain

$$\{\nabla_{\mathbf{M}} \operatorname{tr} \mathbf{W}\mathbf{M}^{-1}\}_{ij} = \frac{\partial \Psi_2[\Psi_1(\mathbf{M})]}{\partial \{\mathbf{M}\}_{ij}} = -\{\mathbf{U}\}_{j.}(\mathbf{U}'\mathbf{M}\mathbf{U})^{-2}\{\mathbf{U}'\}_{.i},$$

which implies Eq. (21). □

IV.2.7. *The L_p-class of Optimality Criteria*

Many optimality criteria are particular cases of a large L_p-class of optimality criteria. (An alternative name for the L_p-class of optimality criteria is the "trace-class", cf. [16].) A criterion belonging to the class is defined by a criterion function of the form

$$\begin{aligned} \phi(\mathbf{M}) &= [m^{-1} \operatorname{tr}(\mathbf{H}\mathbf{M}^{-1}\mathbf{H}')^p]^{1/p} \quad \text{if } \det \mathbf{M} \neq 0, \\ &= \infty \qquad\qquad\qquad\qquad\quad\; \text{if } \det \mathbf{M} = 0, \end{aligned} \tag{22}$$

where $p > 0$ and $\mathbf{H}$ is a nonsingular $m \times m$ matrix. (For the definition of the power of a matrix see the introduction to Chapter IV.)

Linear optimality criteria are obtained from Eq. (22) if $p = 1$. The criterion of E-optimality belongs to the L_p-class as can be shown when taking $\mathbf{H} = \mathbf{I}$ and $p \to \infty$. Indeed, denote the eigenvalues of $\mathbf{M}$ by $\lambda_1, \ldots,$

λ_m and the minimum eigenvalue by λ_{i_0}. We can write

$$\lim_{p\to\infty} (\mathrm{tr}\, \mathbf{M}^{-p})^{1/p} = \lim_{p\to\infty} \left(\sum_{i=1}^{m} \lambda_i^{-p}\right)^{1/p}$$

$$= \lim_{p\to\infty} \lambda_{i_0}^{-1} \left(1 + \sum_{i \neq i_0} \left(\frac{\lambda_i}{\lambda_{i_0}}\right)^{-p}\right)^{1/p} = \lambda_{i_0}^{-1}.$$

In turn, the D-optimality criterion is obtained for $\mathbf{H} = \mathbf{I}$ and $p \to 0$. In fact

$$\lim_{p\to 0} \log \left[\frac{1}{m} \mathrm{tr}\, \mathbf{M}^{-p}\right]^{1/p} = \lim_{p\to 0} \frac{\log \frac{1}{m} \sum_{i=1}^{m} \lambda_i^{-p}}{p} = -\frac{1}{m} \log \det \mathbf{M}.$$

Proposition IV.11. *Let* $\mathbf{A}$, $\mathbf{B}$ *be nonzero positive semidefinite* $m \times m$ *matrices and let* r *be a positive integer. Then*

$$\mathbf{A} > \mathbf{B} \Rightarrow \left[\frac{1}{m} \mathrm{tr}\, \mathbf{A}^r\right]^{1/r} > \left[\frac{1}{m} \mathrm{tr}\, \mathbf{B}^r\right]^{1/r}; \quad (r > 1), \tag{23}$$

$$\frac{1}{m} \mathrm{tr}\, \mathbf{AB} \leqslant \left[\frac{1}{m} \mathrm{tr}\, \mathbf{A}^r\right]^{1/r} \left[\frac{1}{m} \mathrm{tr}\, \mathbf{B}^{r/(r-1)}\right]^{(r-1)/r}; \quad (r > 1), \tag{24}$$

$$\left[\frac{1}{m} \mathrm{tr}\, (\mathbf{A} + \mathbf{B})^r\right]^{1/r} \leqslant \left[\frac{1}{m} \mathrm{tr}\, \mathbf{A}^r\right]^{1/r} + \left[\frac{1}{m} \mathrm{tr}\, \mathbf{B}^r\right]^{1/r}. \tag{25}$$

Proof. Suppose $r > 1$. Denote the $m \times m$ matrix composed of orthonormal eigenvectors of $\mathbf{A}$ by $\mathbf{U}$. We have

$$\mathbf{UU}' = \mathbf{I}, \quad \mathbf{U}'\mathbf{AU} \equiv \Lambda = \text{diagonal}.$$

Denote $\mathbf{C} \equiv \mathbf{U}'\mathbf{BU}$. From $\mathbf{A} > \mathbf{B}$ it follows that

$$\Lambda > \mathbf{C}.$$

Denote the orthonormal eigenvectors and the eigenvalues of $\mathbf{C}$ by $\mathbf{v}^{(1)}, \ldots, \mathbf{v}^{(m)}$ and by $\mu_1, \ldots, \mu_m$. Evidently,

$$\mu_i < \mathbf{v}^{(i)\prime} \Lambda \mathbf{v}^{(i)} = \sum_{k=1}^{m} (v_k^{(i)})^2 \{\Lambda\}_{kk};$$

$$(i = 1, \ldots, m).$$

Thus

$$\operatorname{tr} \mathbf{C}^r = \sum_{i=1}^m \mu_i^r < \sum_{i=1}^m \left[\sum_{k=1}^m (v_k^{(i)})^2 \{\Lambda\}_{kk}\right]^r$$
$$\leq \sum_{i=1}^m \left[\sum_{k=1}^m (v_k^{(i)})^2 \{\Lambda\}_{kk}^r\right] = \sum_{i=1}^m \mathbf{v}^{(i)\prime} \Lambda^r \mathbf{v}^{(i)} = \operatorname{tr} \Lambda^r.$$

We have used the Hölder inequality (cf. [109], Chapter XIV, § 1):

$$\sum_{k=1}^m (v_k^{(i)})^2 \{\Lambda\}_{kk} \leq \left[\sum_{k=1}^m (v_k^{(i)})^2 \{\Lambda\}_{kk}^r\right]^{1/r} \left[\sum_{k=1}^m (v_k^{(i)})^2\right]^{(r-1)/r}.$$

So the implication in Eq. (23) has been proven.

Take $\boldsymbol{0} \neq \boldsymbol{u} \in R^m$ and denote $s = r/(r-1)$. Denote by h_u the function

$$h_u\colon \mathbf{C} \in \mathcal{R}^{m \times m} \mapsto \boldsymbol{u}' \left[\frac{\mathbf{C}^r}{r} + \frac{\mathbf{B}^s}{s} - \mathbf{CB}\right] \boldsymbol{u}.$$

Verify that

$$\nabla_{\mathbf{C}} h_u(\mathbf{C}) = \boldsymbol{uu}'\mathbf{C}^{r-1} - \boldsymbol{uu}'\mathbf{B}.$$

Hence

$$\nabla_{\mathbf{C}} h_u(\mathbf{C} = \mathbf{B}^{1/(r-1)}) = \boldsymbol{0}.$$

The $m^2 \times m^2$ matrix of second-order derivatives of h_u is positive definite, thus h_u achieves its minimum at the point $\mathbf{C} = \mathbf{B}^{1/(r-1)}$ regardless of the choice of the vector $\boldsymbol{u} \in R^m$. It follows that

$$\frac{\mathbf{C}^r}{r} + \frac{\mathbf{B}^s}{s} - \mathbf{CB} \geq \boldsymbol{0}. \tag{26}$$

Set $\mathbf{A}/[m^{-1} \operatorname{tr} \mathbf{A}^r]^{1/r}$ instead of $\mathbf{C}$ and set $\mathbf{B}/[m^{-1} \operatorname{tr} \mathbf{B}^s]^{1/s}$ instead of $\mathbf{B}$ in Eq. (26). Compute the trace of the matrix on the left-hand side of Eq. (26). The obtained inequality implies Eq. (24) after some elementary computation.

Evidently,

$$\frac{1}{m} \operatorname{tr} (\mathbf{A} + \mathbf{B})^r = \frac{1}{m} \operatorname{tr} \mathbf{A}(\mathbf{A} + \mathbf{B})^{r-1}$$
$$+ \frac{1}{m} \operatorname{tr} \mathbf{B}(\mathbf{A} + \mathbf{B})^{r-1}. \tag{27}$$

Setting $(\mathbf{A}+\mathbf{B})^{r-1}$ instead of $\mathbf{B}$ into Eq. (24) we obtain

$$\frac{1}{m}\operatorname{tr}\mathbf{A}(\mathbf{A}+\mathbf{B})^{r-1} \leqslant \left[\frac{1}{m}\operatorname{tr}\mathbf{A}^r\right]^{1/r}\frac{1}{m}[\operatorname{tr}(\mathbf{A}+\mathbf{B})^r]^{1/s}.$$

A similar inequality can be obtained for the second term on the right-hand side of Eq. (27). So we obtain from Eq. (27)

$$\frac{1}{m}\operatorname{tr}(\mathbf{A}+\mathbf{B})^r \leqslant \left\{\left[\frac{1}{m}\operatorname{tr}\mathbf{A}^r\right]^{1/r} + \left[\frac{1}{m}\operatorname{tr}\mathbf{B}^r\right]^{1/r}\left[\frac{1}{m}\operatorname{tr}(\mathbf{A}+\mathbf{B})^r\right]^{1/s}\right\},$$

which implies Eq. (25) for the case $r>1$. The equality in Eq. (25) is evident when $r=1$. □

Proposition IV.12. *If p is a positive integer, then the function ϕ defined by Eq. (22) is*

a) *continuous on $\mathfrak{M}$,*
b) *convex on $\mathfrak{M}$ and strictly convex on $\mathfrak{M}_+$,*
c) *differentiable on $\mathfrak{M}_+$, and*

$$\nabla_{\mathbf{M}}\phi(\mathbf{M}) = -\frac{1}{pm^{1/p}}[\operatorname{tr}(\mathbf{H}\mathbf{M}^{-1}\mathbf{H}')^p]^{(1/p)-1} \times \sum_{h=0}^{p-1}\mathbf{M}^{h-p}\mathbf{H}'^p\mathbf{H}^p\mathbf{M}^{-h-1}. \tag{28}$$

Proof. a) The mapping

$$\mathbf{M}\in\mathfrak{M}_+ \mapsto \mathbf{H}\mathbf{M}^{-1}\mathbf{H}'$$

is continuous (see Proposition IV.4). Furthermore, the mapping

$$\mathbf{A} \mapsto \|\mathbf{A}\|_p$$

is continuous on the set of positive semidefinite matrices, since from Proposition IV. 11 (Eq. (25)) it follows that

$$|\,\|\mathbf{A}\|_p - \|\mathbf{B}\|_p| \leqslant \|\mathbf{A}-\mathbf{B}\|_p.$$

Thus ϕ is continuous on $\mathfrak{M}_+$.

Prove now the continuity of ϕ at $\mathbf{M}(\xi)\in\mathfrak{M}-\mathfrak{M}_+$. Denote by g_i the functional

$$g_i(\vartheta)\equiv\{\mathbf{H}\}_{i.}\boldsymbol{\alpha}; \quad (\boldsymbol{\vartheta}=\boldsymbol{\alpha}'\boldsymbol{f}\in\Theta).$$

From Proposition II.8 it follows that

$$\operatorname{var}_\xi g_{i_0}=\infty$$

for some $i_0\in\{1, \ldots, m\}$. Hence from Proposition III.14 we obtain

$$\lim_{n\to\infty} \operatorname{var}_{\xi_n} g_{i_0}=\infty \tag{29}$$

whenever $\lim_{n\to\infty} \mathbf{M}(\xi_n)=\mathbf{M}(\xi)$. Set $\mathbf{A}=\mathbf{HM}^{-1}(\xi_n)\mathbf{H}'$, $\mathbf{B}=\mathbf{I}$, $r=p$ in the inequality (24). We obtain

$$\|\mathbf{HM}^{-1}(\xi_n)\mathbf{H}'\|_p\geq\frac{1}{m}\operatorname{tr}\mathbf{HM}^{-1}(\xi_n)\mathbf{H}=\frac{1}{m}\sum_{i=1}^{m}\operatorname{var}_{\xi_n} g_i.$$

Hence, according to Eq. (29)

$$\lim_{n\to\infty}\|\mathbf{HM}^{-1}(\xi_n)\mathbf{H}'\|_p=\infty=\phi[\mathbf{M}(\xi)].$$

b) According to Proposition III.13, for every $\beta\in(0, 1)$, $\mathbf{M}_1, \mathbf{M}_2\in\mathfrak{M}_+$, $\boldsymbol{u}\in R^m$, $\boldsymbol{u}\neq 0$ we have

$$\begin{aligned}&\boldsymbol{u}'\mathbf{H}[(1-\beta)\mathbf{M}_1+\beta\mathbf{M}_2]^{-1}\mathbf{H}'\boldsymbol{u}\\ &<(1-\beta)\boldsymbol{u}'\mathbf{HM}_1^{-1}\mathbf{H}'\boldsymbol{u}+\beta\boldsymbol{u}'\mathbf{HM}_2^{-1}\mathbf{H}'\boldsymbol{u}.\end{aligned}$$

Therefore, we can use Eq. (23) when setting the matrices

$$(1-\beta)\mathbf{HM}_1^{-1}\mathbf{H}'+\beta\mathbf{HM}_2^{-1}\mathbf{H}'$$

and

$$\mathbf{H}[(1-\beta)\mathbf{M}_1+\beta\mathbf{M}_2]^{-1}\mathbf{H}'$$

instead of the matrices **A** and **B** and taking $r=p$. We obtain

$$\begin{aligned}&\|\mathbf{H}[(1-\beta)\mathbf{M}_1+\beta\mathbf{M}_2]^{-1}\mathbf{H}'\|_p\\ &<\|(1-\beta)\mathbf{HM}_1^{-1}\mathbf{H}'+\beta\mathbf{HM}_2^{-1}\mathbf{H}'\|_p\end{aligned} \tag{30}$$

(see the definition of the norm $\| \|_p$ in Chapter IV.1). Set $\mathbf{A} = (1-\beta)\mathbf{H}\mathbf{M}_1^{-1}\mathbf{H}'$, $\mathbf{B} = \beta\mathbf{H}\mathbf{M}_2^{-1}\mathbf{H}'$ into Eq. (25) and compare with Eq. (30). We obtain the strict convexity of ϕ on $\mathfrak{M}_+$. The convexity on $\mathfrak{M}$ follows from the fact that $\det \mathbf{M} = 0$ implies $\phi(\mathbf{M}) = \infty$.

c) Denote by $\mathbf{I}^{(i,j)}$ the $m \times m$ matrix

$$\{\mathbf{I}^{(i,j)}\}_{k,l} = 0 \text{ if } i \neq k \text{ or } j \neq l,$$
$$= 1 \text{ if } i = k \text{ and } j = l.$$

By induction with respect to p it can be verified that

$$\frac{\partial \mathbf{A}^{-p}}{\partial \{\mathbf{A}\}_{ij}} = -\sum_{h=0}^{p-1} \mathbf{A}^{-h-1}\mathbf{I}^{(i,j)}\mathbf{A}^{-p+h}$$

for every positive definite matrix $\mathbf{A}$, hence that

$$\nabla_{\mathbf{A}} \operatorname{tr}(\mathbf{W}\mathbf{A}^{-p}) = -\sum_{h=0}^{p-1} \mathbf{A}^{-p+h}\mathbf{W}\mathbf{A}^{-h-1} \tag{31}$$

for every symmetric matrix $\mathbf{W}$. In the particular case of $\mathbf{W} = \mathbf{I}$ we obtain

$$\nabla_{\mathbf{A}} \operatorname{tr} \mathbf{A}^{-p} = -p\mathbf{A}^{-p-1}.$$

Using Eq. (31) we obtain the expression in Eq. (28). □

IV.3. PARTIAL OPTIMALITY CRITERIA

In some experiments the experimenter is interested only in some parameters or some functions of the parameters. In such cases partial optimality criteria are used. Another name used for them is "singular optimality criteria".

An example of a partial optimality criterion is given by the criterion function

$$\phi: \mathbf{M}(\xi) \in \mathfrak{M} \mapsto \operatorname{var}_\xi g.$$

The properties of this function are given in Propositions III.13 and III.14. In contrast to the global criteria, ϕ is not continuous on the whole $\mathfrak{M}$, and moreover, the ϕ-optimum design is singular in some cases. Other partial criteria functions have similar properties.

Some partial criteria functions are defined when using functions of the covariance matrix of the BLUE-s of some (not all) parameters. Without restrictions on generality it will be supposed that these are the first s parameters $\alpha_1, \ldots, \alpha_s$ $(s<m)$. Observe that the parameter α_i is estimable when $\mathbf{e}^{(i)} \in \mathcal{M}(\mathbf{M})$.

Basic optimality criteria are the following:

A) *The criterion of partial (or "restricted") D-optimality.* It is defined by

$$\begin{aligned} \phi_0(\mathbf{M}) &= \log \det [\mathbf{M}^-]_\mathrm{I} \text{ if } \mathbf{e}^{(i)} \in \mathcal{M}(\mathbf{M}); \ (i=1, \ldots, s), \\ &= \infty \qquad \text{otherwise,} \end{aligned} \tag{32}$$

where $\mathbf{M}^-$ is an arbitrary g-inverse of $\mathbf{M}$ and $[\mathbf{M}^-]_\mathrm{I}$ is its $s \times s$ submatrix (see the end of Chapter IV.1). The criterion is interpreted as in Chapter IV.2.1.

B) *The L_p-class of partial optimality criteria. The corresponding criteria functions have the form*

$$\begin{aligned} \phi_p(\mathbf{M}) &= [s^{-1} \operatorname{tr} (\mathbf{H}\mathbf{M}^{-1}\mathbf{H}')^p]^{1/p} \text{ if } \mathcal{M}(\mathbf{H}') \subset \mathcal{M}(\mathbf{M}), \\ &= \infty \qquad \text{if } \mathcal{M}(\mathbf{H}') \not\subset \mathcal{M}(\mathbf{M}), \end{aligned} \tag{33}$$

where p is a positive number and $\mathbf{H}$ is an $s \times m$ matrix of full rank s.

In the particular case of $p=1$ and $\mathbf{W}=\mathbf{H}'\mathbf{H}$ we obtain the partial linear optimality criteria. Particularly when $\mathbf{H}=\mathbf{I}$, the partial A-optimality criterion is obtained.

C) *The product optimality criterion is defined by*

$$\phi_S(\mathbf{M}) = \sum_{i_1, \ldots, i_r=1}^{m} \sum_{j_1, \ldots, j_r=1}^{m} \gamma_{i_1, \ldots, i_r} \prod_{k=1}^{r} \{\mathbf{M}^-\}_{i_k j_k} \gamma_{j_1, \ldots, j_r}, \tag{34}$$

provided $\gamma_{\cdot, i_2, \ldots, i_r} \in \mathcal{M}(\mathbf{M})$ for every $i_2, \ldots, i_r$. Otherwise, the value of ϕ_S is

$$\phi_S(\mathbf{M}) = \infty.$$

The coefficients $\gamma_{i_1, \ldots, i_r}$ are supposed to be symmetric with respect to any permutation of the subscripts $i_1, \ldots, i_r$.

In symbols defined in Chapter IV.1, Eq. (34) can be abbreviated as

$$\phi_S(\mathbf{M}) = \gamma'[\mathbf{M}^-]^{\otimes r}\gamma; \quad (\gamma \in \mathcal{M}[\mathbf{M}^{\otimes r}]).$$

The product optimality criterion is derived from the variance of the best unbiassed estimate of a homogeneous polynomial functional (see Chapter VII and [71], [73]).

D) *Minimax partial optimality criteria* are defined by

$$\phi_E(\mathbf{M}) = \sup_{\boldsymbol{h} \in \mathcal{H}} \boldsymbol{h}'\mathbf{M}^- \boldsymbol{h},$$

where $\mathcal{H}$ is a bounded subset of R^m that does not span R^m. In the particular case of

$$\mathcal{H} = \left\{ \boldsymbol{h}: \boldsymbol{h} = \sum_{i=1}^{s} c_i \boldsymbol{e}^{(i)}, \sum_{i=1}^{s} c_i^2 = 1 \right\},$$

we obtain the partial E-optimality criterion. Notice that $\boldsymbol{e}^{(i)}$ is the i-th column of the identity matrix.

EXERCISE IV.7. Let $\mathbf{W}$ be a positive semidefinite matrix of rank s. Prove that the function

$$\begin{aligned} \phi_{\mathbf{W}}(\mathbf{M}) &= \operatorname{tr} \mathbf{W}\mathbf{M}^- \quad \text{if } \mathcal{M}(\mathbf{W}) \subset \mathcal{M}(\mathbf{M}), \\ &= \infty \qquad\qquad \text{if } \mathcal{M}(\mathbf{W}) \not\subset \mathcal{M}(\mathbf{M}) \end{aligned}$$

is a particular case of the function ϕ_p from Eq. (33).

HINT. Using Proposition III.12 prove that there is an $s \times m$ matrix $\mathbf{H}$ of rank s such that $\mathbf{W} = \mathbf{H}'\mathbf{H}$.

EXERCISE IV.8. Let g be a linear functional defined on Θ. Prove that the optimality criterion defined by

$$\phi[\mathbf{M}(\xi)] = \operatorname{var}_\xi g; \quad (\xi \in \Xi)$$

is a particular case of the criterion (33).

HINT. Take $s = 1$.

EXERCISE IV.9. Let $\mathbf{H}$ be an $s \times m$ matrix of rank s. Prove that the function

$$\begin{aligned} \phi(\mathbf{M}) &= \log \det (\mathbf{H}\mathbf{M}\mathbf{H}')^{-1} \quad \text{if } \mathcal{M}(\mathbf{H}') \subset \mathcal{M}(\mathbf{M}), \\ &= \infty \qquad\qquad\qquad\qquad \text{otherwise} \end{aligned}$$

is the D-optimality criterion function taking another linear basis of Θ instead of $f_1, \ldots, f_m$.

HINT. Show that the condition $\mathcal{M}(\mathbf{H}') \subset \mathcal{M}(\mathbf{M})$ ensures the existence of the matrix $[\mathbf{H}\mathbf{M}\mathbf{H}']^{-1}$. Add suplementary rows to the matrix $\mathbf{H}$ to obtain a nonsingular matrix $\mathbf{T}$. Define a new basis as in Exercise IV.4.

Now prove the following auxiliary statement.

PROPOSITION IV.13 [52]. *Let* $\mathbf{A}$, $\mathbf{B}$ *be positive definite* $m \times m$ *matrices and let* $\mathbf{C}$ *be an* $s \times m$ *matrix of rank* s. *Then*

$$(\mathbf{C}\mathbf{A}^{-1}\mathbf{C}')^{-1} + (\mathbf{C}\mathbf{B}^{-1}\mathbf{C}')^{-1} \leqslant [\mathbf{C}(\mathbf{A}+\mathbf{B})^{-1}\mathbf{C}']^{-1}. \tag{35}$$

PROOF. The inequality (35) is evident if $m = s$. The case of $m > s$ can be proved by induction with respect to m. According to Proposition III.12, there is a nonsingular matrix $\mathbf{U}$ such that $\mathbf{U}'\mathbf{A}^{-1}\mathbf{U} = \mathbf{I}$ and $\mathbf{U}'\mathbf{B}^{-1}\mathbf{U} = \operatorname{diag}(\lambda_1, \ldots, \lambda_m)$. Therefore, suppose in the proof that $\mathbf{A}^{-1} = \mathbf{I}$ and that $\mathbf{B}^{-1} = \operatorname{diag}(\lambda_1, \ldots, \lambda_m)$. In that case

$$\mathbf{C}\mathbf{A}^{-1}\mathbf{C}' = \sum_{i=1}^{m} \{\mathbf{C}\}_{\cdot i}\{\mathbf{C}\}_{i\cdot},$$

$$\mathbf{C}\mathbf{B}^{-1}\mathbf{C}' = \sum_{i=1}^{m} \lambda_i\{\mathbf{C}\}_{\cdot i}\{\mathbf{C}\}_{i\cdot},$$

$$\mathbf{C}(\mathbf{A}+\mathbf{B})^{-1}\mathbf{C}' = \sum_{i=1}^{m} \frac{\lambda_i}{1+\lambda_i}\{\mathbf{C}\}_{\cdot i}\{\mathbf{C}\}_{i\cdot}.$$

Denote the sums on the right-hand sides but for the case that $1 \leqslant i \leqslant m-1$ by $\mathbf{W}$, $\mathbf{V}$, $\mathbf{Q}$, and denote $\mathbf{z} = \{\mathbf{C}\}_{m}$. We have to prove that the inequality

$$\mathbf{W}^{-1} + \mathbf{V}^{-1} \leqslant \mathbf{Q}^{-1}$$

implies

$$(\mathbf{W}+\mathbf{z}\mathbf{z}')^{-1} + (\mathbf{V}+\lambda_m\mathbf{z}\mathbf{z}')^{-1} \leqslant \left(\mathbf{Q} + \frac{\lambda_m}{1+\lambda_m}\mathbf{z}\mathbf{z}'\right)^{-1}. \tag{36}$$

We can verify that

$$(\mathbf{W}+\mathbf{z}\mathbf{z}')^{-1} = \mathbf{W}^{-1} - \frac{\mathbf{W}^{-1}\mathbf{z}\mathbf{z}'\mathbf{W}^{-1}}{1+\mathbf{z}'\mathbf{W}^{-1}\mathbf{z}},$$

after multiplying by the matrix $\mathbf{W}+\mathbf{z}\mathbf{z}'$. We obtain similar expressions for the other terms in Eq. (36). Thus, we have to prove that the matrix

$$\mathbf{P} \equiv \Delta + \frac{\mathbf{W}^{-1}\mathbf{z}\mathbf{z}'\mathbf{W}^{-1}}{1+\mathbf{z}'\mathbf{W}^{-1}\mathbf{z}} + \lambda_m \frac{\mathbf{V}^{-1}\mathbf{z}\mathbf{z}'\mathbf{V}^{-1}}{1+\lambda_m\mathbf{z}'\mathbf{V}^{-1}\mathbf{z}} - \frac{\lambda_m\mathbf{Q}^{-1}\mathbf{z}\mathbf{z}'\mathbf{Q}^{-1}}{1+\lambda_m+\lambda_m\mathbf{z}'\mathbf{Q}^{-1}\mathbf{z}}$$

is positive semidefinite whenever the matrix

$$\Delta \equiv \mathbf{Q}^{-1} - \mathbf{W}^{-1} - \mathbf{V}^{-1} \tag{37}$$

is positive semidefinite. For that purpose we shall use the following inequality between the numbers a, b, c, d, β ($\beta \geqslant 0$, $c > d > 0$):

$$\frac{a^2}{\beta+d} - \frac{b^2}{\beta+c} + \frac{[|b|-|a|]^2}{c-d} \geqslant 0. \tag{38}$$

This inequality follows from the elementary inequality

$$\frac{a^2}{\beta+d} - \frac{b^2}{\beta+c} + \frac{[|b|-|a|]^2}{c-d} = \frac{[|a|(\beta+c)-|b|(\beta+d)]^2}{(\beta+c)(\beta+d)(c-d)} \geqslant 0.$$

Take $\boldsymbol{u} \in R^m$. Using adequate a, b, c, d, β we obtain from Eq. (38) that

$$\begin{aligned}\boldsymbol{u}'\mathbf{P}\boldsymbol{u} &= \boldsymbol{u}'\Delta\boldsymbol{u} + \frac{(\boldsymbol{u}'\mathbf{W}^{-1}\mathbf{z})^2}{1+\mathbf{z}'\mathbf{W}^{-1}\mathbf{z}} + \frac{(\boldsymbol{u}'\mathbf{V}^{-1}\mathbf{z})}{\lambda_m^{-1}+\mathbf{z}'\mathbf{V}^{-1}\mathbf{z}} \\ &\quad - \frac{(\boldsymbol{u}'\mathbf{Q}^{-1}\mathbf{z})^2}{1+\lambda_m^{-1}+\mathbf{z}'\mathbf{Q}^{-1}\mathbf{z}} \geqslant \boldsymbol{u}'\Delta\boldsymbol{u} + \frac{(\boldsymbol{u}'\mathbf{W}^{-1}\mathbf{z})^2}{1+\mathbf{z}'\mathbf{W}^{-1}\mathbf{z}} \\ &\quad - \frac{[|\boldsymbol{u}'\mathbf{Q}^{-1}\mathbf{z}|-|\boldsymbol{u}'\mathbf{V}^{-1}\mathbf{z}|]^2}{1+\mathbf{z}'\mathbf{Q}^{-1}\mathbf{z}-\mathbf{z}'\mathbf{V}^{-1}\mathbf{z}}.\end{aligned}$$

Further, according to Eq. (37)

$$\begin{aligned}&|(|\boldsymbol{u}'\mathbf{Q}^{-1}\mathbf{z}|-|\boldsymbol{u}'\mathbf{V}^{-1}\mathbf{z}|)| \\ &\leqslant |\boldsymbol{u}'(\mathbf{Q}^{-1}-\mathbf{V}^{-1})\mathbf{z}| \leqslant |\boldsymbol{u}'\Delta\mathbf{z}| + |\boldsymbol{u}'\mathbf{W}^{-1}\mathbf{z}|.\end{aligned}$$

Thus

$$\begin{aligned}\boldsymbol{u}'\mathbf{P}\boldsymbol{u} &\geqslant \boldsymbol{u}'\Delta\boldsymbol{u} + \frac{(\boldsymbol{u}'\mathbf{W}^{-1}\mathbf{z})^2}{1+\mathbf{z}'\mathbf{W}^{-1}\mathbf{z}} - \frac{[|\boldsymbol{u}'\Delta\mathbf{z}|+|\boldsymbol{u}'\mathbf{W}^{-1}\mathbf{z}|]^2}{1+\mathbf{z}'\mathbf{W}^{-1}\mathbf{z}+\mathbf{z}'\Delta\mathbf{z}} \\ &\geqslant \boldsymbol{u}'\Delta\boldsymbol{u} - \frac{(\boldsymbol{u}'\Delta\mathbf{z})^2}{\mathbf{z}'\Delta\mathbf{z}} \geqslant 0,\end{aligned} \tag{39}$$

where, besides the Schwarz inequality

$$(\boldsymbol{u}'\Delta\mathbf{z})^2 \leqslant \boldsymbol{u}'\Delta\boldsymbol{u}\mathbf{z}'\Delta\mathbf{z}$$

the inequality (38) has been used again. □

PPOPOSITION IV.14. *The function* ϕ_0 *defined by Eq. (32) is*

a) *convex on* $\mathfrak{M}$,

b) *continuous and differentiable on* $\mathfrak{M}_+$, *and*

$$\nabla_{\mathbf{M}}\phi_0(\mathbf{M}) = -\mathbf{M}^{-1} + \begin{pmatrix} \mathbf{0}, & \mathbf{0} \\ \mathbf{0}, & [\mathbf{M}_{\mathrm{III}}]^{-1} \end{pmatrix}. \tag{40}$$

PROOF. a) Take $\mathbf{M}_1, \mathbf{M}_2 \in \mathfrak{M}_{\phi_0} \equiv \{\mathbf{M} : \mathbf{M} \in \mathfrak{M}, \phi_0(\mathbf{M}) < \infty\}$. Denote by $\mathbf{H}$ the $s \times m$ matrix

$$\{\mathbf{H}\}_{ii} = 1; \quad (i = 1, \ldots, s), \quad \{\mathbf{H}\}_{ij} = 0; \quad (i \neq j).$$

Evidently,

$$[\mathbf{M}_i^+]_{\mathrm{I}} = [\mathbf{M}_i^-]_{\mathrm{I}} = \mathbf{H}\mathbf{M}_i^-\mathbf{H}'; \quad (i = 1, 2)$$

regardless of the choice of the g-inverse $\mathbf{M}_i^-$. Hence, using Propositions IV.13 and II.9, we can write

$$\begin{aligned}
&(1-\beta)\{[\mathbf{M}_1^-]_{\mathrm{I}}\}^{-1} + \beta\{[\mathbf{M}_2^-]_{\mathrm{I}}\}^{-1} \\
&= \lim_{\varepsilon \searrow 0} \{(1-\beta)[\mathbf{H}(\mathbf{M}_1 + \varepsilon\mathbf{I})^{-1}\mathbf{H}']^{-1} + \beta[\mathbf{H}(\mathbf{M}_2 + \varepsilon\mathbf{I})^{-1}\mathbf{H}']^{-1}\} \\
&\leqslant \lim_{\varepsilon \searrow 0} \{\mathbf{H}[(1-\beta)\mathbf{M}_1 + \beta\mathbf{M}_2 + \varepsilon\mathbf{I}]^{-1}\mathbf{H}'\}^{-1} \\
&= \{[((1-\beta)\mathbf{M}_1 + \beta\mathbf{M}_2)^-]_{\mathrm{I}}\}^{-1}.
\end{aligned} \tag{41}$$

From Eq. (41) and the concavity of the function

$$\mathbf{M} \in \mathfrak{M} \mapsto \log \det \mathbf{M}$$

(Proposition IV.2) we obtain

$$\phi_0[(1-\beta)\mathbf{M}_1 + \beta\mathbf{M}_2] \leqslant (1-\beta)\phi_0(\mathbf{M}_1) + \beta\phi_0(\mathbf{M}_2). \tag{42}$$

The inequality (42) is obvious if $\mathbf{M}_1 \notin \mathfrak{M}_{\phi_0}$ or if $\mathbf{M}_2 \notin \mathfrak{M}_{\phi_0}$.

b) From the equality

$$\begin{pmatrix} \mathbf{I}, & \mathbf{0} \\ \mathbf{0}, & \mathbf{I} \end{pmatrix} = \begin{pmatrix} \mathbf{M}_{\mathrm{I}}, & \mathbf{M}_{\mathrm{II}} \\ \mathbf{M}'_{\mathrm{II}}, & \mathbf{M}_{\mathrm{III}} \end{pmatrix} \begin{pmatrix} [\mathbf{M}^{-1}]_{\mathrm{I}}, & [\mathbf{M}^{-1}]_{\mathrm{II}} \\ [\mathbf{M}^{-1}]'_{\mathrm{II}}, & [\mathbf{M}^{-1}]_{\mathrm{III}} \end{pmatrix},$$

after performing the multiplication indicated on the right-hand side, we obtain

$$\mathbf{M}_{I}[\mathbf{M}^{-1}]_{I}+\mathbf{M}_{II}[\mathbf{M}^{-1}]'_{II}=\mathbf{I}, \tag{43}$$

$$\mathbf{M}'_{II}[\mathbf{M}^{-1}]_{I}+\mathbf{M}_{III}[\mathbf{M}^{-1}]'_{II}=\mathbf{0}. \tag{44}$$

From Eq. (43)

$$[(\mathbf{M}^{-1})_{I}]^{-1}=\mathbf{M}_{I}+\mathbf{M}_{II}(\mathbf{M}^{-1})'_{II}[(\mathbf{M}^{-1})_{I}]^{-1},$$

hence after substituting for $(\mathbf{M}^{-1})_{II}$ according to Eq. (44), we obtain

$$[(\mathbf{M}^{-1})_{I}]^{-1}=\mathbf{M}_{I}-\mathbf{M}_{II}(\mathbf{M}_{III})^{-1}\mathbf{M}'_{II}. \tag{45}$$

On the left-hand side of Eq. (45) there is the inverse of the covariance matrix of the BLUE-s for $\alpha_1, \ldots, \alpha_s$, corresponding to the information matrix $\mathbf{M}$. The equality (45) can therefore be useful also for other partial optimality criteria.

Define an $m \times m$ matrix $\mathbf{L}$ by

$$\mathbf{L}=\begin{pmatrix}\mathbf{I}, & -\mathbf{M}_{II}(\mathbf{M}_{III})^{-1}\\ \mathbf{0}, & \mathbf{I}\end{pmatrix}.$$

Evidently, $\det \mathbf{L}=1$ and

$$\mathbf{LM}=\begin{pmatrix}\mathbf{M}_{I}-\mathbf{M}_{II}(\mathbf{M}_{III})^{-1}\mathbf{M}'_{II}, & \mathbf{0}\\ \mathbf{M}'_{II}, & \mathbf{M}_{III}\end{pmatrix}.$$

Therefore, using Eq. (45), we obtain

$$\det \mathbf{M}=\det \mathbf{L} \det \mathbf{M}=\det \mathbf{M}_{III} \det [(\mathbf{M}^{-1})_{I}]^{-1}.$$

If $\mathbf{M}=\mathbf{M}(\xi)$, then

$$\phi[\mathbf{M}(\xi)]=-\log \det \mathbf{M}(\xi)+\log \det \mathbf{M}_{III}(\xi). \tag{46}$$

Hence the continuity of ϕ at nonsingular information matrices is a consequence of the part a) of Proposition IV.2. Similarly, Eq. (40) is obtained from Eq. (46) and from the part c) of the same proposition. □

PROPOSITION IV.15. *In the case of p being a positive integer, the function ϕ_p defined in Eq. (33) is*

a) *convex on* $\mathfrak{M}$,
b) *continuous and differentiable on* $\mathfrak{M}_+$, *and*

$$\nabla_{\mathbf{M}}\phi_p(\mathbf{M}) = -\frac{1}{ps^{1/p}}[\operatorname{tr}(\mathbf{HM}^{-1}\mathbf{H}')^p]^{(1/p)-1}\sum_{h=0}^{p-1}\mathbf{M}^{h-p}(\mathbf{H}'\mathbf{H})^p\mathbf{M}^{-h-1}. \quad (47)$$

PROOF. a) Let $\mathbf{M}_1, \mathbf{M}_2 \in \mathfrak{M}_{\phi_p}$ and $\beta \in (0, 1)$. The matrix $(\mathbf{M}_1 + \varepsilon\mathbf{I})$ is positive definite when $\varepsilon > 0$, hence, according to Proposition III.12, there is a nonsingular matrix $\mathbf{U}$ such that

$$\mathbf{U}'(\mathbf{M}_1 + \varepsilon\mathbf{I})\mathbf{U} = \mathbf{I}, \quad \mathbf{U}'\mathbf{M}_2\mathbf{U} = \mathbf{\Lambda} = \text{diagonal}.$$

From the convexity of the function $x \in (0, \infty) \mapsto x^{-1}$ it follows that for every $\boldsymbol{z} \in R^m$

$$\begin{aligned}
&\boldsymbol{z}'\{\mathbf{H}[(1-\beta)(\mathbf{M}_1+\varepsilon\mathbf{I})+\beta\mathbf{M}_2]^{-1}\mathbf{H}'\}\boldsymbol{z} \\
&= (\mathbf{U}'\mathbf{H}'\boldsymbol{z})'[(1-\beta)\mathbf{I}+\beta\mathbf{\Lambda}]^{-1}(\mathbf{U}'\mathbf{H}'\boldsymbol{z}) \\
&\leqslant (\mathbf{U}'\mathbf{H}'\boldsymbol{z})'[(1-\beta)\mathbf{I}+\beta\mathbf{\Lambda}^{+}](\mathbf{U}'\mathbf{H}'\boldsymbol{z}) \\
&= \boldsymbol{z}'[(1-\beta)\mathbf{H}(\mathbf{M}_1+\varepsilon\mathbf{I})^{-1}\mathbf{H}'+\beta\mathbf{HM}_2^{-}\mathbf{H}']\boldsymbol{z}.
\end{aligned}$$

Taking the limit for $\varepsilon \searrow 0$ and using Proposition II.9 we obtain that

$$(1-\beta)\mathbf{HM}_1^{-}\mathbf{H}'+\beta\mathbf{HM}_2^{-}\mathbf{H}' \geqslant \mathbf{H}[(1-\beta)\mathbf{M}_1+\beta\mathbf{M}_2]^{-}\mathbf{H}'. \quad (48)$$

By a modification of the proof of Eq. (23) we can prove that

$$\begin{aligned}
&\|(1-\beta)\mathbf{HM}_1^{-}\mathbf{H}'+\beta\mathbf{HM}_2^{-}\mathbf{H}'\|_p \\
&\geqslant \|\mathbf{H}[(1-\beta)\mathbf{M}_1+\beta\mathbf{M}_2]^{-}\mathbf{H}'\|_p.
\end{aligned}$$

From Eq. (25) it follows that

$$\begin{aligned}
&\|(1-\beta)\mathbf{HM}_1^{-}\mathbf{H}'+\beta\mathbf{HM}_2^{-}\mathbf{H}'\|_p \\
&\leqslant (1-\beta)\|\mathbf{HM}_1^{-}\mathbf{H}'\|_p+\beta\|\mathbf{HM}_2^{-}\mathbf{H}'\|_p.
\end{aligned}$$

The last two inequalities imply

$$\phi_p[(1-\beta)\mathbf{M}_1+\beta\mathbf{M}_2] \leqslant (1-\beta)\phi_p(\mathbf{M}_1)+\beta\phi_p(\mathbf{M}_2).$$

A similar inequality is obvious when $\mathbf{M}_1 \notin \mathfrak{M}_{\phi_p}$ or $\mathbf{M}_2 \notin \mathfrak{M}_{\phi_p}$.

b) We prove the continuity of ϕ_p and we verify the expression in Eq. (47) as in Proposition IV.12. □

EXERCISE IV.10. Verify that

$$\nabla_{\mathbf{M}} \operatorname{tr} \mathbf{W}\mathbf{M}^{-1} = -\mathbf{M}^{-1}\mathbf{W}\mathbf{M}^{-1}$$

and that

$$\nabla_{\mathbf{M}} \mathbf{g}'\mathbf{M}^{-1}\mathbf{g} = -\mathbf{M}^{-1}\mathbf{g}\mathbf{g}'\mathbf{M}^{-1}.$$

PROPOSITION IV.16. *The function ϕ_S defined in Eq. (34) is*

a) *convex on $\mathfrak{M}$,*

b) *continuous and differentiable on $\mathfrak{M}_+$, and*

$$\nabla_{\mathbf{M}}\phi_S(\mathbf{M}) = -r \sum_{i_1,\ldots,i_r=1}^{m} \sum_{j_1,\ldots,j_r=1}^{m} \gamma_{i_1,\ldots,i_r} \{\mathbf{M}^{-1}\}_{\cdot i_1} \times \prod_{k=2}^{r} \{\mathbf{M}^{-1}\}_{i_k j_k} \{\mathbf{M}^{-1}\}_{j_1 \cdot} \gamma_{j_1,\ldots,j_r}. \tag{49}$$

PROOF. a) We shall use the abbreviated notation

$$\phi_S(\mathbf{M}) = \gamma'[\mathbf{M}^-]^{\otimes r}\gamma; \quad (\text{if } \phi_S(\mathbf{M}) < \infty).$$

To prove the convexity of ϕ_S it is sufficient to verify the inequalities

$$\gamma'\{[(1-\beta)\mathbf{M}_1 + \beta\mathbf{M}_2]^-\}^{\otimes r}\gamma \leq \gamma[(1-\beta)\mathbf{M}_1^- + \beta\mathbf{M}_2^-]^{\otimes r}\gamma, \tag{50}$$

$$\gamma'[(1-\beta)\mathbf{M}_1^- + \beta\mathbf{M}_2^-]^{\otimes r}\gamma \leq (1-\beta)\gamma'[\mathbf{M}_1^-]^{\otimes r}\gamma + \beta\gamma'[\mathbf{M}_2^-]^{\otimes r}\gamma, \tag{51}$$

as follows:

Denote by Ψ the function

$$\Psi\colon \beta \in \langle 0, 1\rangle \mapsto \gamma'[(1-\beta)\mathbf{M}_1^- + \beta\mathbf{M}_2^-]^{\otimes r}\gamma.$$

Exploiting the symmetry of the coefficients $\gamma_{i_1,\ldots,i_r}$ we obtain

$$\frac{d^2\Psi(\beta)}{d\beta^2} = r(r-1)\gamma'[\mathbf{M}_2^- - \mathbf{M}_1^-]$$
$$\otimes[(1-\beta)\mathbf{M}_1^- + \beta\mathbf{M}_2^-]^{\otimes(r-2)} \otimes (\mathbf{M}_2^- - \mathbf{M}_1^-)\gamma \geq 0.$$

Hence Ψ is convex, which means that the inequality (51) holds.

From Proposition II.13 it follows that

$$\boldsymbol{g}'[(1-\beta)\mathbf{M}_1 + \beta\mathbf{M}_2]^- \boldsymbol{g} \leq (1-\beta)\boldsymbol{g}'\mathbf{M}_1^- \boldsymbol{g} + \beta \boldsymbol{g}'\mathbf{M}_2^- \boldsymbol{g}$$

for every vector $\boldsymbol{g} \in \mathcal{M}(\mathbf{M}_1) \cap \mathcal{M}(\mathbf{M}_2)$. This allows us to prove Eq. (50) as follows:

Let $\mathbf{Q}$ be the $m \times m$ matrix

$$\{\mathbf{Q}\}_{i_1 j_1} = \sum_{i_2, \ldots, i_r = 1}^{m} \sum_{j_2, \ldots, j_r = 1}^{m} \gamma_{i_1, \ldots, i_r} \prod_{k=2}^{r} \{[(1-\beta)\mathbf{M}_1 + \beta\mathbf{M}_2]^-\}_{i_k j_k} \, \gamma_{j_1, \ldots, j_r}. \tag{52}$$

The matrix $\mathbf{Q}$ is positive semidefinite. Denote the eigenvalues by $\lambda_1, \ldots, \lambda_m$, and the orthonormal eigenvectors of $\mathbf{Q}$ by $\boldsymbol{u}_1, \ldots, \boldsymbol{u}_m$. Let $\mathbf{U} \equiv (\boldsymbol{u}_1, \ldots, \boldsymbol{u}_m)$ and $\boldsymbol{\Lambda} \equiv \operatorname{diag}(\lambda_1, \ldots, \lambda_m)$. Obviously,

$$\mathbf{Q} = \mathbf{U}\boldsymbol{\Lambda}\mathbf{U}' = \sum_{i=1}^{m} \lambda_i \boldsymbol{u}_i \boldsymbol{u}_i'. \tag{53}$$

Therefore

$$\gamma'\{[(1-\beta)\mathbf{M}_1 + \beta\mathbf{M}_2]^-\}^{\otimes r}\gamma = \operatorname{tr}[(1-\beta)\mathbf{M}_1 + \beta\mathbf{M}_2]^- \mathbf{Q}$$
$$= \sum_{i=1}^{m} \lambda_i \boldsymbol{u}_i'[(1-\beta)\mathbf{M}_1 + \beta\mathbf{M}_2]^- \boldsymbol{u}_i$$
$$\leq \sum_{i=1}^{m} \lambda_i \boldsymbol{u}_i'[(1-\beta)\mathbf{M}_1^- + \beta\mathbf{M}_2^-]\boldsymbol{u}_i$$
$$= \operatorname{tr}[(1-\beta)\mathbf{M}_1^- + \beta\mathbf{M}_2^-]\mathbf{Q}$$
$$= \gamma'[(1-\beta)\mathbf{M}_1^- + \beta\mathbf{M}_2^-] \otimes \{[(1-\beta)\mathbf{M}_1 + \beta\mathbf{M}_2]^-\}^{\otimes(r-1)}\gamma.$$

Choosing a new matrix $\mathbf{Q}$ we similarly decompose the tensor power

$$\{[(1-\beta)\mathbf{M}_1 + \beta\mathbf{M}_2]^-\}^{\otimes(r-1)}$$

etc., until we obtain Eq. (50).

b) Let $\mathbf{M}_1, \mathbf{M}_2 \in \mathfrak{M}_+$ and define $\mathbf{Q}_1, \mathbf{Q}_2$ like $\mathbf{Q}$ in Eq. (52), setting $\mathbf{M}_1$

or $\mathbf{M}_2$ instead of $(1-\beta)\mathbf{M}_1+\beta\mathbf{M}_2$. We can write

$$\begin{aligned}
&|\phi_S(\mathbf{M}_1)-\phi_S(\mathbf{M}_2)| = |\operatorname{tr} \mathbf{M}_1^{-1}\mathbf{Q}_1 - \operatorname{tr} \mathbf{M}_2^{-1}\mathbf{Q}_2| \\
&\leq \|\mathbf{M}_1^{-1}-\mathbf{M}_2^{-1}\| \, \|\mathbf{Q}_1-\mathbf{Q}_2\| \\
&\leq \ldots \leq \|\mathbf{M}_1^{-1}-\mathbf{M}_2^{-1}\|^r \sup \{|\gamma_{i_1,\ldots,i_r}|^2 : i_k \in \{1,\ldots,m\}\}.
\end{aligned}$$

Thus the continuity of ϕ_S follows from Proposition IV.4. From Eq. (53) and from the symmetry of the coefficients $\gamma_{i_1,\ldots,i_r}$ it follows that

$$\begin{aligned}
\{\nabla_{\mathbf{M}}\phi_S(\mathbf{M})\}_{ij} &= r\left\{\nabla_{\mathbf{M}} \sum_{k=1}^{m} \lambda_k \mathbf{u}_k' \mathbf{M}^{-1} \mathbf{u}_k\right\}_{ij} \\
&= r \sum_{k=1}^{m} \lambda_k \mathbf{u}_k' \left[-\mathbf{M}^{-1} \frac{\partial \mathbf{M}}{\partial \{\mathbf{M}\}_{ij}} \mathbf{M}^{-1}\right] \mathbf{u}_k \\
&= -r \operatorname{tr} \mathbf{Q}\mathbf{M}^{-1}\mathbf{I}^{(i,j)}\mathbf{M}^{-1} \\
&= -r\{\mathbf{M}^{-1}\mathbf{Q}\mathbf{M}^{-1}\}_{ij}. \ \square
\end{aligned}$$

PROPOSITION IV.17. *The function $\phi_E(\mathbf{M})$ is convex on $\mathfrak{M}$ and continuous on $\mathfrak{M}_+$.*

The proof is similar to the proof of Proposition IV.8. □

IV.4. PROPERTIES OF CONVEX CRITERIA FUNCTIONS

A commom property of all criteria functions presented in Chapters IV.2 and IV.3 is their convexity. We consider some consequences of convexity in this chapter.

The functions $\phi(\mathbf{M}) = -\log\det\mathbf{M}$, $\phi(\mathbf{M}) = \operatorname{tr}\mathbf{M}^{-1}$ etc., considered in Chapter IV.2, are continuous on $\mathfrak{M}$. We shall show that continuity is a property of any function ϕ corresponding to a global optimality criterion assuming that

(A) There is a set $\mathcal{U}_\phi \subset \mathcal{L}(\mathfrak{M})$ such that

a) $\mathfrak{M}_+ \subset \mathcal{U}_\phi$,

b) $\mathcal{U}_\phi$ is open relatively to $\mathcal{L}(\mathfrak{M})$,

c) ϕ is defined, finite and convex on $\mathcal{U}_\phi$.

(B) If $\mathbf{M}_n \in \mathfrak{M}_+$; $(n=1,\ldots,\ldots)$ and

$$\lim_{n\to\infty} \mathbf{M}_n = \mathbf{M} \in \mathfrak{M} - \mathfrak{M}_+,$$

then

$$\lim_{n\to\infty} \phi(\mathbf{M}_n) = \infty.$$

REMARKS. The set $\mathcal{U}_\phi$ need not be convex. The function ϕ is only expected to be convex on any convex subset of $\mathcal{U}_\phi$.

We recall that $\mathcal{L}(\mathfrak{M})$ is the linear space of (symmetric) matrices spanned by the set $\mathfrak{M}$.

EXERCISE IV.11. Let $\mathcal{U}_\phi$ be the set of all positive definite matrices belonging to $\mathcal{L}(\mathfrak{M})$. Verify that (A) and (B) hold for the function

$$\mathbf{L} \in \mathcal{L}(\mathfrak{M}) \mapsto -\log \det \mathbf{L}.$$

HINT. Prove the convexity of ϕ on $\mathcal{U}_\phi$ as in the proof of Proposition IV.2.

PROPOSITION IV.18. *If $\mathcal{U}$ is an open subset of $\mathcal{L}(\mathfrak{M})$ and if ϕ is convex and finite on $\mathcal{U}$, then ϕ is continuous on $\mathcal{U}$.*

PROOF [104]. Denote the dimension of $\mathcal{L}(\mathfrak{M})$ by r. Choose $\mathbf{L}_1 \in \mathcal{U}$. Since $\mathcal{U}$ is open, there is an r-dimensional open sphere

$$\mathcal{G} \equiv \{\mathbf{L} \colon \mathbf{L} \in \mathcal{L}(\mathfrak{M}),\ \|\mathbf{L} - \mathbf{L}_1\| < \varrho\}$$

and an r-dimensional simplex $\mathcal{T}$ so that $\mathcal{G} \subset \mathcal{T} \subset \mathcal{U}$. Denote the vertices of the simplex $\mathcal{T}$ by $\mathbf{V}_1, \ldots, \mathbf{V}_r$. From the convexity of ϕ it follows that

$$\phi(\mathbf{L}) \leqslant \max\{\phi(\mathbf{V}_i) \colon i = 0, \ldots, r\} \equiv d < \infty; \quad (\mathbf{L} \in \mathcal{G}).$$

Take $\varepsilon \in (0, 1)$ and denote

$$\mathcal{G}_\varepsilon \equiv \{\mathbf{L} \colon \mathbf{L} \in \mathcal{L}(\mathfrak{M}),\ \|\mathbf{L} - \mathbf{L}_1\| < \varepsilon\varrho\}.$$

For $\mathbf{L} \in \mathcal{G}_\varepsilon$ define $\mathbf{W} \equiv (\mathbf{L} - \mathbf{L}_1)/\varepsilon$. Thus

$$\mathbf{L} = (1 - \varepsilon)\mathbf{L}_1 + \varepsilon(\mathbf{L}_1 + \mathbf{W})$$

and the convexity of ϕ implies

$$\phi(\mathbf{L}) \leqslant (1 - \varepsilon)\phi(\mathbf{L}_1) + \varepsilon d. \tag{54}$$

On the other hand, we have the obvious equality

$$\mathbf{L}_1 = \frac{1}{1+\varepsilon}\,\mathbf{L} + \left(1 - \frac{1}{1+\varepsilon}\right)(\mathbf{L}_1 - \mathbf{W}).$$

Hence, using the convexity of ϕ again, we obtain

$$\phi(\mathbf{L}_1) \leqslant \frac{1}{1+\varepsilon}\,\phi(\mathbf{L}) + \frac{\varepsilon d}{1+\varepsilon}. \tag{55}$$

From Eqs. (54) and (55) it follows that for every $\varepsilon \in (0, 1)$

$$|\phi(\mathbf{L}) - \phi(\mathbf{L}_1)| \leqslant \varepsilon[d - \phi(\mathbf{L}_1)]$$

whenever $\mathbf{L} \in \mathscr{G}_\varepsilon$. The required continuity of ϕ at $\mathbf{L}_1$ follows immediately. □

PROPOSITION IV.19. *Any global criterion function ϕ satisfying the assumptions* (A) *and* (B) *is continuous on* $\mathfrak{M}$.

PROOF. Proposition IV.18 implies the continuity on $\mathfrak{M}_+$, the assumption (B) implies the continuity on $\mathfrak{M} - \mathfrak{M}_+$. □

Partial criteria functions may have discontinuities on $\mathfrak{M}$, as demonstrated in the example in Chapter III.3. Nevertheless, owing to the compactness of the set $\mathfrak{M}$ (Proposition III.10) the lower semicontinuity of ϕ on $\mathfrak{M}$ is sufficient to ensure the existence of a ϕ-optimum design (see the remarks preceding Proposition III.14).

EXERCISE IV.12. Verify that the function

$$\begin{aligned} f(x) &= x^{-1}; \quad 0.1 \leqslant x < 1 \\ &= 5; \qquad x = 1 \end{aligned}$$

is convex but it does not achieve its minimum on the interval $0.1 \leqslant x \leqslant 1$.

We present some necessary and sufficient conditions for the lower semicontinuity of ϕ.

PROPOSITION IV.20 (cf. [104]). *The set* (*the epigraph of the function* ϕ)

$$\operatorname{epi}(\phi) \equiv \{(\mathbf{M}, t) : \mathbf{M} \in \mathfrak{M},\ t \in R,\ t \geqslant \phi(\mathbf{M})\} \tag{56}$$

is closed exactly if ϕ is lower semicontinuous on $\mathfrak{M}$.

PROOF. ϕ is lower semicontinuous at $\mathbf{M}$ iff $\mathbf{M}_n \to \mathbf{M}$ implies

$$\liminf_{n\to\infty} \phi(\mathbf{M}_n) \geqslant \phi(\mathbf{M}). \tag{57}$$

Suppose that Eq. (57) holds and that $t_n \to t$, $\mathbf{M}_n \to \mathbf{M}$, $\phi(\mathbf{M}_n) \leqslant t_n$. Then obviously $t \geqslant \phi(\mathbf{M})$ and the set epi (ϕ) is closed.

Conversely, suppose that epi (ϕ) is closed and $\mathbf{M}_n \to \mathbf{M}$. Denote $c \equiv \liminf_{n\to\infty} \phi(\mathbf{M}_n)$. Hence there is a sequence $\{\mathbf{M}_{n_k}\}_k$ such that

$$\phi(\mathbf{M}_{n_k}) \leqslant c + \frac{1}{k}; \quad (k = 1, 2, \ldots).$$

We can write

$$\phi(\mathbf{M}) = \phi(\lim_k \mathbf{M}_{n_k}) \leqslant \lim_k \left(c + \frac{1}{k}\right) = \liminf_{n\to\infty} \phi(\mathbf{M}_n),$$

thus ϕ is lower semicontinuous. □

To prove Proposition IV.22, a hyperplane tangent to the set epi (ϕ) is needed. Without proof we state the following, intuitively clear proposition. An elementary proof is given in [88], Section 1 d. 2. A detailed analysis of the properties of tangent and separating hyperplanes is in [103], Chapter III. § 11.

PROPOSITION IV.21. *If C is a convex subset of R^k and if $\mathbf{u} \in R^k$ is a boundary point of C, then there is a tangent hyperplane containing the point $\mathbf{u}$, i.e. there is $\mathbf{h} \in R^k$, $\mathbf{h} \neq \mathbf{0}$ such that*

$$\inf \{\mathbf{h}'\mathbf{z}: \mathbf{z} \in C\} = \mathbf{h}'\mathbf{u}.$$

PROPOSITION IV.22 [104]. *The function ϕ is convex and lower semicontinuous exactly if there is a class of functions*

$$\{\phi_j: j \in J\}$$

that are continuous and affine on $\mathfrak{M}$, and such that

$$\phi_j(\mathbf{M}) \leqslant \phi(\mathbf{M}); \quad (j \in J, \mathbf{M} \in \mathfrak{M}),$$
$$\phi(\mathbf{M}) = \sup \{\phi_j(\mathbf{M}): j \in J\}.$$

PROOF. The function $\phi(\mathbf{M}) \equiv \sup \{\phi_j(\mathbf{M}): j \in J\}$ is obviously convex on $\mathfrak{M}$. Suppose that $\mathbf{M}_n \to \mathbf{M}$. For every $j \in J$

$$\liminf_{n\to\infty} \phi(\mathbf{M}_n) \geq \lim_{n\to\infty} \phi_j(\mathbf{M}_n) = \phi_j(\mathbf{M}).$$

It follows that

$$\liminf_{n\to\infty} \phi(\mathbf{M}_n) \geq \sup \{\phi_j(\mathbf{M}): j \in J\} = \phi(\mathbf{M}).$$

Hence ϕ is lower semicontinuous at $\mathbf{M}$.

Conversely, let ϕ be convex and lower semicontinuous. The set epi (ϕ) is convex, and according to Proposition IV.20 it is closed. Hence, according to Proposition IV.21 there is a hyperplane in R^{m+1} which is tangent to epi (ϕ) at the boundary point $(\mathbf{M}, \phi(\mathbf{M}))$. That means, there is a linear, continuous functional $\phi_{\mathbf{M}}$ such that $\phi_{\mathbf{M}}(\mathbf{M}) = \phi(\mathbf{M})$ and

$$\phi_{\mathbf{M}}(\mathbf{M}^*) \leq \phi(\mathbf{M}^*); \quad (\mathbf{M}^* \in \mathfrak{M}).$$

Obviously,

$$\phi(\mathbf{M}^*) = \sup \{\phi_{\mathbf{M}}(\mathbf{M}^*): \mathbf{M} \in \mathfrak{M},$$
$$\mathbf{M} \text{ is a boundary point of } \mathfrak{M}\}; \ (\mathbf{M}^* \in \mathfrak{M}). \ \square$$

EXERCISE IV.13. Prove Proposition III.14 using Propositions IV.22, IV.4 and Eq. (15) in Chapter II.

PROPOSITION IV.23 [104]. *Let ϕ be convex on $\mathfrak{M}$ and let $\mathbf{M}_1, \mathbf{M}_2 \in \mathfrak{M}$. The function*

$$\varphi: \beta \in (0, 1) \mapsto \frac{\phi[(1-\beta)\mathbf{M}_1 + \beta\mathbf{M}_2] - \phi(\mathbf{M}_1)}{\beta}$$

is nondecreasing on (0, 1), hence there always is a finite or infinite $(= -\infty)$ limit

$$\lim_{\beta \searrow 0} \frac{\phi[(1-\beta)\mathbf{M}_1 + \beta\mathbf{M}_2] - \phi(\mathbf{M}_1)}{\beta}. \tag{58}$$

PROOF. Suppose that $0 < \beta_1 < \beta_2 < 1$. From the convexity of ϕ it follows that

$$\phi[(1-\beta_1)\mathbf{M}_1 + \beta_1\mathbf{M}_2] - \phi[\mathbf{M}_1]$$

$$=\phi\left\{\frac{\beta_1}{\beta_2}[(1-\beta_2)\mathbf{M}_1+\beta_2\mathbf{M}_2]+\left(1-\frac{\beta_1}{\beta_2}\right)\mathbf{M}_1\right\}-\phi[\mathbf{M}_1]$$
$$\leq\frac{\beta_1}{\beta_2}\{\phi[(1-\beta_2)\mathbf{M}_1+\beta_2\mathbf{M}_2]-\phi[\mathbf{M}_1]\},$$

which implies the required monotonicity. □

The limit in Eq. (58) will be denoted by $\partial\phi(\mathbf{M}_1, \mathbf{M}_2)$. It is called the (Frechét) directional derivative of ϕ at the point $\mathbf{M}_1$ in the direction of $\mathbf{M}_2$.

PROPOSITION IV.24. *Assume that the criterion function ϕ has the property (A). Then for every $\mathbf{L}_1\in\mathcal{U}_\phi$, $\mathbf{L}_2\in\mathcal{U}_\phi$ we have*

$$\partial\phi(\mathbf{L}_1, \mathbf{L}_2)>-\infty \tag{59}$$

and the function

$$\mathbf{L}\in\mathcal{U}_\phi\mapsto\partial\phi(\mathbf{L}_1, \mathbf{L})$$

is convex and continuous.

PROOF. From (A) it follows that there is an open sphere

$$\{\mathbf{L}: \mathbf{L}\in\mathcal{L}(\mathfrak{M}), \|\mathbf{L}-\mathbf{L}_1\|<\varrho\}\subset\mathcal{U}_\phi$$

of the same dimension as $\mathcal{L}(\mathfrak{M})$. It follows that there is a $\delta>0$ so that the intercept of the straight line

$$\{(1-\beta)\mathbf{L}_1+\beta\mathbf{L}_2: \beta\in\langle 0, \delta\rangle\}$$

is a subset of $\mathcal{U}_\phi$, and hence the function ϕ is finite and convex on it.

Therefore

$$\phi(\mathbf{L}_1)=\phi\left\{\frac{\delta}{\beta+\delta}[(1-\beta)\mathbf{L}_1+\beta\mathbf{L}_2]+\frac{\beta}{\beta+\delta}[(1+\delta)\mathbf{L}_1-\delta\mathbf{L}_2]\right\}$$
$$\leq\frac{\delta}{\beta+\delta}\phi[(1-\beta)\mathbf{L}_1+\beta\mathbf{L}_2]+\frac{\beta}{\beta+\delta}\phi[(1+\delta)\mathbf{L}_1-\delta\mathbf{L}_2].$$

After some rearrangements we obtain

$$\frac{\phi[(1-\beta)\mathbf{L}_1+\beta\mathbf{L}_2]-\phi[\mathbf{L}_1]}{\beta} \geq \frac{\phi[\mathbf{L}_1]-\phi[(1+\delta)\mathbf{L}_1-\delta\mathbf{L}_2]}{\delta};$$

$(\beta \in (0, \delta))$,

which implies, after taking the limit for $\beta \searrow 0$, that $\partial\phi(\mathbf{L}_1, \mathbf{L}_2) > -\infty$.

For every fixed $\beta > 0$ the function

$$\mathbf{L} \in \mathcal{U}_\phi \mapsto \frac{\phi[(1-\beta)\mathbf{L}_1+\beta\mathbf{L}]-\phi(\mathbf{L}_1)}{\beta}$$

is convex. It follows that the limit for $\beta \searrow 0$, i.e. the function

$$\mathbf{L} \in \mathcal{U}_\phi \mapsto \partial\phi(\mathbf{L}_1, \mathbf{L})$$

is convex as well, and according to Eq. (59) it is finite on the open set $\mathcal{U}_\phi$. From Proposition IV.18 it therefore follows that ϕ is continuous on $\mathcal{U}_\phi$. □

From the definition of $\nabla_\mathbf{M}\phi(\mathbf{M})$ (cf. Chapter IV.1) it follows that

$$\begin{aligned}\partial\phi(\mathbf{M}, \mathbf{L}) &= \operatorname{tr} \nabla_\mathbf{M}\phi(\mathbf{M})(\mathbf{L}-\mathbf{M}) \\ &= \langle \nabla_\mathbf{M}\phi(\mathbf{M}), \mathbf{L}-\mathbf{M} \rangle; \quad (\mathbf{L} \in \mathcal{R}^{m \times m}).\end{aligned} \tag{60}$$

The converse statement holds as well ([103]): If $\mathbf{V} \in \mathcal{R}^{m \times m}$ is such that

$$\partial\phi(\mathbf{M}, \mathbf{L}) = \langle \mathbf{V}, \mathbf{L}-\mathbf{M} \rangle; \quad (\mathbf{L} \in \mathcal{R}^{m \times m}),$$

then there exists $\nabla_\mathbf{M}\phi(\mathbf{M})$ and it is equal to $\mathbf{V}$.

If $\mathbf{M} \in \mathcal{L} \equiv$ a linear subspace of $\mathcal{R}^{m \times m}$ and if $\mathbf{V} \in \mathcal{L}$ such that

$$\partial\phi(\mathbf{M}, \mathbf{L}) = \langle \mathbf{V}, \mathbf{L}-\mathbf{M} \rangle; \quad (\mathbf{L} \in \mathcal{L}), \tag{61}$$

then $\mathbf{V}$ is the gradient of ϕ "relative to $\mathcal{L}$" and it will be denoted by

$$\nabla_\mathbf{M}\phi(\mathbf{M} \mid \mathcal{L}).$$

From Eq. (61) it follows that $\nabla_\mathbf{M}\phi(\mathbf{M} \mid \mathcal{L})$ is obtained by projecting the matrix $\nabla_\mathbf{M}\phi(\mathbf{M})$ onto $\mathcal{L}$.

Without proving we present the following proposition (for the proof cf. [103], Corollary 25.5.1).

PROPOSITION IV.25. *If ϕ is finite and convex on an open subset $\mathcal{U}$ of $\mathcal{L}$*

and if $\nabla_{\mathbf{L}}\phi(\mathbf{L}|\mathscr{L})$ *is defined at any* $\mathbf{L}\in\mathscr{L}$, *then the mapping*

$$\mathbf{L}\in\mathscr{U}\mapsto\nabla_{\mathbf{L}}\phi(\mathbf{L}|\mathscr{L})$$

is continuous.

DEFINITION IV.2. *A design* ξ^* *is said to be locally* ϕ*-optimum if for every* $\xi\in\Xi$ *the function*

$$\varphi: \beta\in\langle 0, 1)\mapsto\phi[(1-\beta)\mathbf{M}(\xi^*)+\beta\mathbf{M}(\xi)]$$

has a local minimum at $\beta=0$.

PROPOSITION IV.26 [16, 56]. *(The general equivalence theorem.) Suppose that* $\partial\phi[\mathbf{M}(\xi^*), \mathbf{M}(\xi)]>-\infty$ *for every* $\xi\in\Xi$. *Then the following properties of* ξ^* *are equivalent*

1) ξ^* *is* ϕ*-optimum,*
2) ξ^* *is locally* ϕ*-optimum,*
3) $\partial\phi[\mathbf{M}(\xi^*), \mathbf{M}(\xi)]\geqslant 0;\quad (\xi\in\Xi)$.

PROOF. The implication $1\Rightarrow 2$ is obvious. If ξ^* is locally ϕ-optimum, then for every $\xi\in\Xi$ there is $\delta>0$ such that

$$\frac{\phi[(1-\beta)\mathbf{M}(\xi^*)+\beta\mathbf{M}(\xi)]-\phi[\mathbf{M}(\xi^*)]}{\beta}\geqslant 0$$

for every $\beta\in(0, \delta)$. It follows that $\partial\phi[\mathbf{M}(\xi^*), \mathbf{M}(\xi)]\geqslant 0$, i.e. $2\Rightarrow 3$ has been proven.

To prove $3\Rightarrow 1$ we suppose that ξ^* is not ϕ-optimum. Then there is a design ξ such that $\phi[\mathbf{M}(\xi)]<\phi[\mathbf{M}(\xi^*)]$. Hence for every $\beta\in(0, 1)$

$$\begin{aligned}&\frac{\phi[(1-\beta)\mathbf{M}(\xi^*)+\beta\mathbf{M}(\xi)]-\phi[\mathbf{M}(\xi^*)]}{\beta}\\ &\leqslant\frac{(1-\beta)\phi[\mathbf{M}(\xi^*)]+\beta\phi[\mathbf{M}(\xi)]-\phi[\mathbf{M}(\xi^*)]}{\beta}\\ &=\phi[\mathbf{M}(\xi)]-\phi[\mathbf{M}(\xi^*)],\end{aligned}$$

thus $\partial\phi[\mathbf{M}(\xi^*), \mathbf{M}(\xi)]\leqslant\phi[\mathbf{M}(\xi)]-\phi[\mathbf{M}(\xi^*)]<0$, and so the inequality in 3) does not hold. □

In case when there is the gradient $\nabla\phi[\mathbf{M}(\xi^*)]$, Proposition IV.26 leads to the following proposition.

PROPOSITION IV.27. *Suppose that the gradient of ϕ is defined at $\mathbf{M}(\xi^*)$. Then the following properties of ξ^* are equivalent:*

1) *ξ^* is ϕ-optimum,*

2) $$\boldsymbol{f}'(x)\nabla\phi[\mathbf{M}(\xi^*)]\boldsymbol{f}(x) \geq \operatorname{tr} \mathbf{M}(\xi^*)\nabla\phi[\mathbf{M}(\xi^*)]; \quad (x \in X), \tag{62}$$

3) $$\min_{\bar{x}\in X} \boldsymbol{f}'(\bar{x})\nabla\phi[\mathbf{M}(\xi^*)]\boldsymbol{f}(\bar{x}) = \sum_{x\in X} \boldsymbol{f}'(x)\nabla\phi[\mathbf{M}(\xi^*)]\boldsymbol{f}(x)\xi^*(x). \tag{63}$$

PROOF. According to Eq. (60)

$$\partial\phi[\mathbf{M}(\xi^*), \boldsymbol{f}(x)\boldsymbol{f}'(x)] = \boldsymbol{f}'(x)\nabla\phi[\mathbf{M}(\xi^*)]\boldsymbol{f}(x) - \operatorname{tr} \mathbf{M}(\xi^*)\nabla\phi[\mathbf{M}(\xi^*)].$$

Therefore the implication $1 \Rightarrow 2$ is a consequence of Proposition IV.25 since $\boldsymbol{f}(x)\boldsymbol{f}'(x) = \mathbf{M}(\xi_x)$. If, in turn, the inequality (62) is multiplied by $\xi(x)$, and the result is summed over X, then

$$\partial\phi[\mathbf{M}(\xi^*), \mathbf{M}(\xi)] \geq 0; \quad (\xi \in \Xi).$$

Hence, according to Proposition IV.25, ξ^* is ϕ-optimum. The implication $3 \Rightarrow 2$ follows clearly from

$$\operatorname{tr} \mathbf{M}(\xi^*)\nabla\phi[\mathbf{M}(\xi^*)] = \sum_{x\in X} \operatorname{tr} \boldsymbol{f}(x)\boldsymbol{f}'(x)\nabla\phi[\mathbf{M}(\xi^*)]\xi^*(x).$$

Conversely, from 2) and from the inequality

$$\min_{\bar{x}\in X} \boldsymbol{f}'(\bar{x})\nabla\phi[\mathbf{M}(\xi^*)]\boldsymbol{f}(\bar{x}) \leq \sum_{x\in X} \boldsymbol{f}'(x)\nabla\phi[\mathbf{M}(\xi^*)]\boldsymbol{f}(x)\xi^*(x)$$

we obtain the statement in 3). □

EXERCISE IV.14. Prove the equivalence of D-optimum and G-optimum designs (Proposition IV.6) from Proposition IV.27.

When computing ϕ-optimum designs it is often necessary to evaluate the difference $\phi[\mathbf{M}(\mu)] - \inf_{\xi} \phi[\mathbf{M}(\xi)]$ for a given design μ.

PROPOSITION IV.28 [52]. *Suppose that* $\phi[\mathbf{M}(\mu)] < \infty$, $\delta > 0$ *and that*

$$\partial\phi[\mathbf{M}(\mu), \mathbf{M}] \geq -\delta; \quad (\mathbf{M} \in \mathfrak{M}). \tag{64}$$

Then

$$\phi[\mathbf{M}(\mu)] \leq \inf\{\phi[\mathbf{M}(\xi)] : \xi \in \Xi\} + \delta. \tag{65}$$

In the case when there is the gradient $\nabla\phi[\mathbf{M}(\mu)]$, *the assumption* (64) *is equivalent to*

$$\begin{aligned} &\mathbf{f}'(x)\nabla\phi[\mathbf{M}(\mu)]\mathbf{f}(x) \\ &\geq \operatorname{tr} \nabla\phi[\mathbf{M}(\mu)]\mathbf{M}(\mu) - \delta; \quad (x \in X). \end{aligned} \tag{66}$$

PROOF. From the convexity of ϕ and from Proposition IV.23 we obtain

$$\begin{aligned} &\phi[\mathbf{M}(\xi)] - \phi[\mathbf{M}(\mu)] \\ &\geq \frac{\phi[(1-\beta)\mathbf{M}(\mu) + \beta\mathbf{M}(\xi)] - \phi[\mathbf{M}(\mu)]}{\beta} \\ &\geq \partial\phi[\mathbf{M}(\mu), \mathbf{M}(\xi)] \geq -\delta. \end{aligned}$$

Hence

$$\phi[\mathbf{M}(\mu)] \leq \phi[\mathbf{M}(\xi)] + \delta; \quad (\xi \in \Xi),$$

which implies the inequality (65).

The equivalence of (64) and (66) follows from the expression for the directional derivative in Eq. (60). □

For nondifferentiable but convex criteria functions we can still use the directional derivative instead of the gradient when testing the optimality of a design. In the following proposition we express this derivative for the E-optimality criterion (cf. [16]), and so we complete the exposition of Chapter IV.2.5.

By $\lambda(\mathbf{A})$ we denote the minimum eigenvalue of a matrix $\mathbf{A}$.

PROPOSITION IV.29. *Let* ϕ *be defined on* $\mathfrak{M}_+$ *by*

$$\phi(\mathbf{M}) = \lambda^{-1}(\mathbf{M})$$

(see Eq. (17)). Let $\mathbf{Q_M}$ *be the matrix with columns equal to all orthonormal eigenvectors of* $\mathbf{M}$ *corresponding to* $\lambda(\mathbf{M})$. *Then*

$$\partial\phi(\mathbf{M},\mathbf{L}) = -\frac{\lambda[\mathbf{Q}'_\mathbf{M}(\mathbf{L}-\mathbf{M})\mathbf{Q}_\mathbf{M}]}{\lambda^2(\mathbf{M})}; \quad (\mathbf{L}\in\mathfrak{M}). \tag{67}$$

In the particular case when $\mathbf{Q_M}$ *is a vector, the function* ϕ *is differentiable at* $\mathbf{M}$ *and*

$$\nabla_\mathbf{M}\phi(\mathbf{M}) = -\frac{\mathbf{Q}_\mathbf{M}\mathbf{Q}'_\mathbf{M}}{\lambda^2(\mathbf{M})}. \tag{68}$$

PROOF. Denote by $\delta(\varepsilon)$ the difference

$$\delta(\varepsilon) \equiv \lambda[(1-\varepsilon)\mathbf{M}+\varepsilon\mathbf{L}] - \lambda(\mathbf{M}).$$

According to Proposition IV.8, $\delta(\cdot)$ is a continuous function of ε. Denote

$$\mathbf{Z}_\varepsilon \equiv (1-\varepsilon)\mathbf{M}+\varepsilon\mathbf{L} - (\lambda(\mathbf{M})+\delta(\varepsilon))\mathbf{I}$$

and denote by

$$\mathbf{U} = \begin{pmatrix}\mathbf{Q}'_\mathbf{M}\\ \mathbf{T}'\end{pmatrix}$$

the $m\times m$ matrix with rows equal to the orthonormal eigenvectors of $\mathbf{M}$. Obviously, $\mathbf{UU}'=\mathbf{I}$ hence $\det\mathbf{U}=1$. Furthermore, $\det\mathbf{Z}_\varepsilon=0$ since $\lambda(\mathbf{M})+\delta(\varepsilon)$ is an eigenvalue of $(1-\varepsilon)\mathbf{M}+\varepsilon\mathbf{L}$. Thus

$$0 = \det \mathbf{UZ}_\varepsilon\mathbf{U}' = \det\begin{pmatrix}\mathbf{Q}'_\mathbf{M}\mathbf{Z}_\varepsilon\mathbf{Q}_\mathbf{M}, & \mathbf{Q}'_\mathbf{M}\mathbf{Z}_\varepsilon\mathbf{T}\\ \mathbf{T}'\mathbf{Z}_\varepsilon\mathbf{Q}_\mathbf{M}, & \mathbf{T}'\mathbf{Z}_\varepsilon\mathbf{T}\end{pmatrix}. \tag{69}$$

For a sufficiently small ε the matrix $\mathbf{T}'\mathbf{Z}_\varepsilon\mathbf{T}$ is nonsingular. Multiply the submatrix $(\mathbf{T}'\mathbf{Z}_\varepsilon\mathbf{Q}_\mathbf{M}, \mathbf{T}'\mathbf{Z}_\varepsilon\mathbf{T})$ by $-(\mathbf{T}'\mathbf{Z}_\varepsilon\mathbf{T})^{-1}\mathbf{Q}'_\mathbf{M}\mathbf{Z}_\varepsilon\mathbf{T}$ and add the results to the submatrix $(\mathbf{Q}'_\mathbf{M}\mathbf{Z}_\varepsilon\mathbf{Q}_\mathbf{M}, \mathbf{Q}'_\mathbf{M}\mathbf{Z}_\varepsilon\mathbf{T})$. From Eq. (69) we obtain

$$\begin{aligned}0 &= \det\mathbf{Z}_\varepsilon\\ &= \det(\mathbf{T}'\mathbf{Z}_\varepsilon\mathbf{T})\det[\mathbf{Q}'_\mathbf{M}\mathbf{Z}_\varepsilon\mathbf{Q}_\mathbf{M} - \mathbf{Q}'_\mathbf{M}\mathbf{Z}_\varepsilon\mathbf{T}(\mathbf{T}'\mathbf{Z}_\varepsilon\mathbf{T})^{-1}\mathbf{T}'\mathbf{Z}_\varepsilon\mathbf{Q}_\mathbf{M}].\end{aligned}$$

It follows that

$$\begin{aligned}0 = \det[&\mathbf{Q}'_\mathbf{M}\mathbf{Z}_\varepsilon\mathbf{Q}_\mathbf{M}\\ &-\varepsilon^2\mathbf{Q}'_\mathbf{M}(\mathbf{L}-\mathbf{M})\mathbf{T}(\mathbf{T}'\mathbf{Z}_\varepsilon\mathbf{T})^{-1}\mathbf{T}'(\mathbf{L}-\mathbf{M})\mathbf{Q}_\mathbf{M}],\end{aligned}$$

where we used the orthogonality of $\mathbf{Q_M}$ and $\mathbf{T}$. Expressing $\mathbf{Q'_M Z_\varepsilon Q_M}$ we obtain

$$0=\det[\varepsilon \mathbf{Q'_M}(\mathbf{L}-\mathbf{M})\mathbf{Q_M}-(\delta(\varepsilon)+\mathcal{O}(\varepsilon^2))\mathbf{I}].$$

It follows that $\delta(\varepsilon)+\mathcal{O}(\varepsilon^2)$ is equal to an eigenvalue of the matrix $\varepsilon\mathbf{Q'_M}(\mathbf{L}-\mathbf{M})\mathbf{Q_M}$ (and it is the minimum eigenvalue for a small ε because of the definition of $\delta(\varepsilon)$). Hence

$$\lim_{\varepsilon\searrow 0}\frac{\delta(\varepsilon)}{\varepsilon}=\lambda[\mathbf{Q'_M}(\mathbf{L}-\mathbf{M})\mathbf{Q_M}].$$

Thus Eq. (67) follows immediately.

In particular, if $\mathbf{Q_M}$ is a vector, then

$$\partial\phi(\mathbf{M},\mathbf{L})=\operatorname{tr}\left(-\frac{\mathbf{Q_M Q'_M}}{\lambda^2(\mathbf{M})}\right)(\mathbf{L}-\mathbf{M}),$$

hence, as explained after Eq. (60), $-\mathbf{Q_M Q'_M}/\lambda^2(\mathbf{M})$ is the gradient of ϕ. □

We investigated Schur's ordering of designs in Propositions III.4 and III.5. We shall prove now that designs which are better according to Schur's ordering are also better with respect to a class of optimality criteria. (Compare with the L_p-class in Chapter IV.2.7.)

We recall that ξ is better than η or equivalent to η, according to Schur's ordering, if for every $k=1,\ldots,m$ there hold the inequalities

$$\sum_{i=1}^{k}\lambda_i(\xi)\geqslant\sum_{i=1}^{k}\lambda_i(\eta).$$

Here $\lambda_1(\xi)\leqslant\lambda_2(\xi)\leqslant\ldots\leqslant\lambda_m(\xi)$ are the ordered eigenvalues of $\mathbf{M}(\xi)$.

PROPOSITION IV.30. *If ξ is better than η or equivalent to η according to Schur's ordering, then for every $p>0$*

$$[\operatorname{tr}\mathbf{M}^{-p}(\xi)]^{1/p}\leqslant[\operatorname{tr}\mathbf{M}^{-p}(\eta)]^{1/p} \tag{70}$$

and this inequality holds also for both $p\to 0$ (=D-optimality), and $p\to\infty$ (=E-optimality).

PROOF. The inequality (70) follows directly from the definition of Schur's ordering in case of E-optimality (see Eq. (17)). In case of D-optimality we can write

$$\det \mathbf{M} = \prod_{i=1}^{m} \lambda_i = v_1[v_2 - v_1] \ldots [v_m - v_{m-1}],$$

where $v_k \equiv \sum_{i=1}^{k} \lambda_i$. It follows that

$$\frac{\partial \log \det \mathbf{M}}{\partial v_i} = \frac{1}{v_i - v_{i-1}} - \frac{1}{v_{i+1} - v_i}$$
$$= \frac{\lambda_{i+1} - \lambda_i}{\lambda_i \lambda_{i+1}} \geqslant 0; \quad (i = 1, \ldots, m).$$

Hence $[-\log \det \mathbf{M}]$ is a decreasing function of the variables $v_1, \ldots, v_m$ which determine Schur's ordering. So Eq. (70) holds for the D-optimality.

Now consider the case of $p \in (0, \infty)$. Let $\mathbf{U} \equiv (\mathbf{u}_1, \ldots, \mathbf{u}_m)$ be the matrix of orthonormal eigenvectors of $\mathbf{M}$. Then

$$\mathbf{U}'\mathbf{M}\mathbf{U} = \operatorname{diag}(\lambda_1, \ldots, \lambda_m),$$

hence

$$\operatorname{tr} \mathbf{M}^{-p} = \operatorname{tr} \mathbf{M}^{-p}\mathbf{U}\mathbf{U}' = \operatorname{tr} \mathbf{U}'\mathbf{M}^{-p}\mathbf{U} = \sum_{i=1}^{m} \lambda_i^{-p}.$$

It follows that

$$[\operatorname{tr} \mathbf{M}^{-p}]^{1/p} = \left[\sum_{i=1}^{m} (v_i - v_{i-1})^{-p}\right]^{1/p}.$$

Thus

$$\frac{\partial [\operatorname{tr} \mathbf{M}^{-p}]^{1/p}}{\partial v_i} = [\operatorname{tr} \mathbf{M}^{-p}]^{(1/p)-1}(\lambda_{i+1}^{-p-1} - \lambda_i^{-p-1}) \leqslant 0;$$
$$(i = 1, \ldots, m).$$

So Eq. (70) is proved. □

IV.5. SINGULAR ϕ-OPTIMUM DESIGNS

Even when a partial criterion function ϕ is differentiable in $\mathfrak{M}_+$ (as in the cases considered in Chapter IV.3), there is no gradient of ϕ at singular information matrices. Therefore we cannot apply Proposition IV.27 or Eq. (66) for checking ϕ-optimum designs. Moreover, as pointed out in the example in Chapter III.3, partial criteria functions can have essential discontinuities at optimum information matrices if the matrices are singular (cf. also [66]). Therefore, singular designs (i.e. designs with singular information matrices) deserve special attention.

Consider criteria functions of the form

$$\begin{aligned} \phi(\mathbf{M}) &= \Psi(\mathbf{H}\mathbf{M}^{-}\mathbf{H}') \quad \text{if } \mathcal{M}(\mathbf{H}') \subset \mathcal{M}(\mathbf{M}), \\ &= \infty \qquad\qquad\qquad \text{otherwise}, \end{aligned} \tag{71}$$

where $\mathbf{H}$ is an $s \times m$ matrix of rank s and Ψ is a convex differentiable function defined on the set of all positive definite $s \times s$ matrices. In accordance with Eq. (1) we shall suppose that Ψ is "nondecreasing", i.e. $\mathbf{A} \geqslant \mathbf{B} \Rightarrow \Psi(\mathbf{A}) \geqslant \Psi(\mathbf{B})$. The convexity of ϕ follows, since

$$\mathbf{H}[(1-\beta)\mathbf{M}^{-} + \beta\mathbf{N}^{-}]\mathbf{H}' \leqslant (1-\beta)\mathbf{H}\mathbf{M}^{-}\mathbf{H}' + \beta\mathbf{H}\mathbf{N}^{-}\mathbf{H}'$$

(see Proposition III.13 with $\boldsymbol{g} = \mathbf{H}'\boldsymbol{h}$ for some $\boldsymbol{h} \in R^m$). The L_p-class and the D-optimality criteria functions are examples of such ϕ.

A generalization of Proposition IV.27 follows.

PROPOSITION IV.31 [12]. *A singular design ξ is ϕ-optimum exactly if there is an $m \times (m$ — rank $\mathbf{M}(\xi))$ matrix $\mathbf{G}$ such that*

(i) *the span of $\mathcal{M}[\mathbf{M}(\xi)] \cup \mathcal{M}(\mathbf{G})$ is the whole space R^m,*

(ii) $$\min_{x \in X} \boldsymbol{f}'(x)\nabla\phi[\mathbf{M}(\xi) + \mathbf{G}\mathbf{G}']\boldsymbol{f}(x) = \operatorname{tr}[\mathbf{M}(\xi) + \mathbf{G}\mathbf{G}']\nabla\phi[\mathbf{M}(\xi) + \mathbf{G}\mathbf{G}'].$$

The sufficiency has been proven [12] and the necessity in [54], using more general convex tools (subgradients). Another approach to a generalization of Propositions IV.26 and IV.27 is in [60].

REMARK. The proof of the sufficiency of the condition (ii) in Proposition IV.31 exploits the fact that $(\mathbf{M}(\xi) + \mathbf{G}\mathbf{G}')^{-1}$ is a nonsingular g-inverse of $\mathbf{M}(\xi)$. It is simple to prove the sufficiency of a condition similar to (ii) for any g-inverse of $\mathbf{M}(\xi)$. However, for most choices of the g-inverses this condition is empty (it never holds).

Discontinuities of the partial criteria functions at optimum designs disappear when using ridge estimates instead of the BLUE-s (see Chapter II.4.D). We recall that a ridge estimate of a linear functional g under a design ξ is proportional to

$$\hat{g}(k, \xi) \equiv \mathbf{g}'\mathbf{M}_k^{-1}(\xi) \sum_{x \in X} \mathbf{f}(x)\xi(x)\bar{y}(x),$$

where

$$\mathbf{M}_k(\xi) \equiv \mathbf{M}(\xi) + k\mathbf{I}$$

and $\bar{y}(x)$ is the arithmetic mean of results of observations performed at x. The variance of $\hat{g}(k, \xi)$ is

$$\text{var}_{\xi, k} g \equiv \mathbf{g}'\mathbf{M}_k^{-1}(\xi)\mathbf{M}(\xi)\mathbf{M}_k^{-1}(\xi)\mathbf{g}$$

and its bias is

$$\text{bias}_{\xi, k, \alpha} g \equiv \mathbf{g}'[\mathbf{M}_k^{-1}(\xi)\mathbf{M}(\xi) - \mathbf{I}]\boldsymbol{\alpha},$$

where $\boldsymbol{\alpha}$ is the true (but unknown) vector of parameters. Thus

$$\begin{aligned} &\text{var}_{\xi, k} g + \text{bias}^2_{\xi, k, \alpha} g \\ &= \mathbf{g}'\mathbf{M}_k^{-1}(\xi)[\mathbf{M}(\xi) + k^2\boldsymbol{\alpha}\boldsymbol{\alpha}']\mathbf{M}_k^{-1}(\xi)\mathbf{g}. \end{aligned} \tag{72}$$

In the case of $k \leq 1/\|\boldsymbol{\alpha}\|^2$ we have

$$\begin{aligned} &\mathbf{u}'(k^2\boldsymbol{\alpha}\boldsymbol{\alpha}')\mathbf{u} = k^2(\mathbf{u}'\boldsymbol{\alpha})^2 \\ &\leq k(\mathbf{u}'\boldsymbol{\alpha})^2/\boldsymbol{\alpha}'\boldsymbol{\alpha} \leq k\mathbf{u}'\mathbf{u} = \mathbf{u}'(k\mathbf{I})\mathbf{u}; \quad (\mathbf{u} \in R^m), \end{aligned}$$

hence

$$\mathbf{M}(\xi) + k^2\boldsymbol{\alpha}\boldsymbol{\alpha}' \leq \mathbf{M}_k(\xi).$$

It follows that

$$\begin{aligned} &\mathbf{g}'\mathbf{M}_k^{-1}(\xi)\mathbf{M}(\xi)\mathbf{M}_k^{-1}(\xi)\mathbf{g} \\ &\leq \text{var}_{\xi, k} g + \text{bias}^2_{\xi, k, \alpha} g \leq \mathbf{g}'\mathbf{M}_k^{-1}(\xi)\mathbf{g}; \\ &(k \leq 1/\|\boldsymbol{\alpha}\|^2, \mathbf{g} \in \mathcal{M}[\mathbf{M}(\xi)]). \end{aligned} \tag{73}$$

From Eq. (72) we obtain (when writing $\mathbf{M} = \mathbf{M}(\xi)$, $\mathbf{M}_k = \mathbf{M}_k(\xi)$)

$$\lim_{k \searrow 0} \mathbf{g}'\mathbf{M}_k^{-1}[\mathbf{M} + k^2\boldsymbol{\alpha}\boldsymbol{\alpha}']\mathbf{M}_k^{-1}\mathbf{g} = \mathrm{var}_\xi g = \mathbf{g}'\mathbf{M}^-\mathbf{g}.$$

Hence, setting $\mathbf{M}_k^{-1}[\mathbf{M} + k^2\boldsymbol{\alpha}\boldsymbol{\alpha}']\mathbf{M}_k^{-1}$ instead of $\mathbf{M}^-$ into Eq. (71), the criterion function corresponding to the ridge estimation is obtained:

$$\phi_{\boldsymbol{\alpha}}^{(k)}(\mathbf{M}) \equiv \Psi[\mathbf{H}\mathbf{M}_k^{-1}(\mathbf{M} + k^2\boldsymbol{\alpha}\boldsymbol{\alpha}')\mathbf{M}_k^{-1}\mathbf{H}'].$$

Here $\phi_{\boldsymbol{\alpha}}^{(0)}(\mathbf{M}) \equiv \phi(\mathbf{M})$; $(\boldsymbol{\alpha} \in R^m)$. Particularly, the function

$$\phi_0^{(k)}(\mathbf{M}) = \Psi[\mathbf{H}\mathbf{M}_k^{-1}\mathbf{M}\mathbf{M}_k^{-1}\mathbf{H}']$$

corresponds to considering the variances but not the bias of ridge estimates. From Eq. (73) it follows that

$$\phi_0^{(k)}(\mathbf{M}) \leq \phi_{\boldsymbol{\alpha}}^{(k)}(\mathbf{M}) \leq \phi(\mathbf{M}); \qquad (74)$$
$$(k < \|\boldsymbol{\alpha}\|^{-2}, \mathbf{0} \neq \mathbf{M} \in \mathfrak{M}).$$

In the following proposition we shall assume that $\|\mathbf{f}(x)\| > c$ for some $c > 0$ and for every $x \in X$. Since the trial x with $\mathbf{f}(x) = 0$ contains no information about the parameters, the assumption is not a relevant lack of generality (compare with Proposition III.7). A consequence of the assumption is that $\mathbf{M} \neq 0$ for every $\mathbf{M} \in \mathfrak{M}$.

Denote by $\mu_k(\boldsymbol{\alpha})$ the $\phi_{\boldsymbol{\alpha}}^{(k)}$-optimum design, i.e.

$$\phi_{\boldsymbol{\alpha}}^{(k)}[\mathbf{M}(\mu_k(\alpha))] = \min \{\phi_{\boldsymbol{\alpha}}^{(k)}(\mathbf{M}) : \mathbf{M} \in \mathfrak{M}\}.$$

PROPOSITION IV.32. *For every* $\boldsymbol{\alpha} \in R^m$

$$\lim_{k \searrow 0} \phi_{\boldsymbol{\alpha}}^{(k)}[\mathbf{M}(\mu_k(\boldsymbol{\alpha}))] = \min \{\phi(\mathbf{M}) : \mathbf{M} \in \mathfrak{M}\} \qquad (75)$$

and

$$\lim_{k \searrow 0} \phi[\mathbf{M}(\mu_k(\boldsymbol{\alpha}))] = \min \{\phi(\mathbf{M}) : \mathbf{M} \in \mathfrak{M}\}. \qquad (76)$$

All limit points of the sequence $\{\mathbf{M}(\mu_{1/n}(\boldsymbol{\alpha}))\}_{n=1}^{\infty}$ *are* ϕ*-optimum information matrices.*

PROOF. According to Eq. (74) it is sufficient to prove the case of $\boldsymbol{\alpha} = 0$. We abbreviate $\mu_k(0)$ by μ_k, and $\phi_0^{(k)}$ by $\phi^{(k)}$. From the monotonicity of Ψ it follows that the function $k \in (0, 1) \mapsto \phi^{(k)}(\mathbf{M})$ is nonincreasing, hence $k \in (0, 1) \mapsto \phi^{(k)}[\mathbf{M}(\mu_k)]$ is nonincreasing as well. Therefore, using Eq. (74)

$$\lim_{k \searrow 0} \phi^{(k)}[\mathbf{M}(\mu_k)] \equiv c \leqslant \min_{\mathbf{M}} \phi(\mathbf{M}). \tag{77}$$

Let $\mathbf{M}^*$ be a limit point of the sequence $\{\mathbf{M}(\mu_{1/n}(\boldsymbol{\alpha}))\}_{n=1}^{\infty}$. Let $\{\mathbf{M}(\mu_{t(n)})\}$ be its subsequence tending to $\mathbf{M}^*$. From the monotonicity of Ψ it follows that for every $h \in (0, \|\boldsymbol{\alpha}\|^{-2})$ and every sufficiently large n

$$\phi^{(h)}[\mathbf{M}(\mu_{t(n)})] \leqslant \phi^{t(n)}[\mathbf{M}(\mu_{t(n)})].$$

Hence

$$\phi^{(h)}[\mathbf{M}^*] = \lim_{n \to \infty} \phi^{(h)}[\mathbf{M}(\mu_{t(n)})] \leqslant c,$$

thus

$$\phi[\mathbf{M}^*] = \lim_{h \searrow 0} \phi^{(h)}[\mathbf{M}^*] \leqslant c.$$

Comparing it with Eq. (77), we obtain

$$\phi(\mathbf{M}^*) = c = \min_{\mathbf{M}} \phi(\mathbf{M}). \ \Box$$

NOTES. For $k > 0$ the function $\mathbf{M} \in \mathfrak{M} \mapsto \phi_{\alpha}^{(k)}(\mathbf{M})$ is continuous and differentiable, though not convex. So, applying both Propositions IV.31 and IV.32, we come to the conclusion that optimum designs are to be checked according to the condition (ii) in Proposition IV.31, but estimation is to be carried out with ridge estimates. Proposition IV.32 guarantees that the ϕ-optimum design is nearly optimal for the ridge estimates as well.

We shall end the chapter by considering the extremal case of all designs being singular, i.e.

$$\mathbf{M}(\xi) = \sum_x \mathbf{f}(x)\mathbf{f}'(x)\xi(x)$$

is singular for every design ξ. It follows that the set $\{\mathbf{f}(x) : x \in X\}$ does not span $\boldsymbol{R}^m$. In fact, it is the case when $f_1(\cdot), \ldots, f_m(\cdot)$ are linearly dependent on X, i.e. they do not constitute a basis of the set of states Θ (Chapter II). This "pathology" could be eliminated by a suitable reparametrization (i.e. by choosing of a new basis in Θ). Nevertheless for various reasons it would be better to retain the original parametrization (see Exercise II.8).

Consider the following "complementarization" of the model. Let $r \equiv \dim \{\mathbf{f}(x): x \in X\}$. Take vectors $\mathbf{f}(z_1), \ldots, \mathbf{f}(z_{m-r}) \in R^m$ which are mutually orthogonal and also orthogonal to $\{\mathbf{f}(x): x \in X\}$. Add "the labels $z_1, \ldots, z_{m-r}$" to the set X,

$$\tilde{X} \equiv X \cup \{z_1, \ldots, z_{m-r}\}.$$

Denote

$$\tilde{\mathbf{M}}(\xi) \equiv \sum_{x \in X} \mathbf{f}(x)\mathbf{f}'(x)\xi(x) + \sum_{i=1}^{m-r} \mathbf{f}(z_i)\mathbf{f}'(z_i)\xi(z_i)$$

for any design ξ supported by a subset of $\tilde{X}$. Denote by

$$\xi_X(x) \equiv \xi(x)/\xi(X); \quad (x \in X)$$

the restriction of ξ to the set X.

Suppose that ϕ has the form (71) with $\mathcal{M}(\mathbf{H}')$ being a subset of the span of $\{\mathbf{f}(x): x \in X\}$.

Proposition IV.33 [67]. *If*

$$\phi[\tilde{\mathbf{M}}(\xi^*)] = \min \{\phi[\tilde{\mathbf{M}}(\xi)]; \xi \text{ on } \tilde{X}\},$$

then ξ^* is supported by a subset of X, and

$$\phi[\mathbf{M}(\xi^*)] = \min \{\phi[\mathbf{M}(\mu)]: \mu \text{ on } X\}.$$

Proof. Applying Proposition III.15 $(m-r)$ times to the matrix

$$\tilde{\mathbf{M}}(\xi) = \xi(X)\mathbf{M}(\xi_X) + \sum_{i=1}^{m-r} \mathbf{f}(z_i)\mathbf{f}'(z_i)\xi(z_i)$$

we obtain

$$\mathbf{g}'\tilde{\mathbf{M}}^+(\xi)\mathbf{g} = \frac{\mathbf{g}'\tilde{\mathbf{M}}^+(\xi_X)\mathbf{g}}{\xi(X)} \geq \mathbf{g}'\tilde{\mathbf{M}}^+(\xi_X)\mathbf{g}$$

for every $\mathbf{g} \in \operatorname{span} \{\mathbf{f}(x): x \in X\}$. Hence

$$\phi[\tilde{\mathbf{M}}(\xi_X)] \leqslant \phi[\tilde{\mathbf{M}}(\xi)].$$

Thus $\xi^* = \xi^*_{\tilde{X}}$, and

$$\begin{aligned}\phi[\mathbf{M}(\xi^*)] &= \min\{\phi[\tilde{\mathbf{M}}(\xi)] : \xi \text{ on } \tilde{X}\} \\ &\leqslant \min\{\phi[\tilde{\mathbf{M}}(\xi)] : \xi \text{ on } X\} \\ &= \min\{\phi[\mathbf{M}(\xi)] : \xi \text{ on } X\}. \ \square\end{aligned}$$

Chapter V

Iterative Computation of Optimum Designs

V.1. INTRODUCTION

To apply the design theory in a real experiment usually means to compute an optimum or almost optimum design. Methods of computation of such designs are considered in this chapter.

We recall that, as in the previous two chapters, the regression model with uncorrelated observations $(X, \Theta, 1)$ is considered, $f_1(\cdot), \ldots, f_m(\cdot)$ being the basis of Θ. The assumptions that X is compact and that $f_1(\cdot), \ldots, f_m(\cdot)$ are continuous on X are important in this chapter.

There are different methods for computing ϕ-optimum designs, especially for the case of D-optimality (see Chapter VI). Some of them allow to obtain the optimum design directly, e.g. using the approximation theory, the game theory, etc. However, they can only be applied to very special cases (e.g. to a polynomial regression on the real line) and they fail to solve general situations. Sufficiently general are only iterative methods, based principally on the following procedure:

a) A starting design is chosen. In principle, it can be any design ξ_0 such that $\det \mathbf{M}(\xi_0) \neq 0$.

b) A sequence of designs $\xi_1, \xi_2, \ldots$ is computed iteratively, the design ξ_{n+1} being obtained by a small perturbation of the design ξ_n, still requiring that $\det \mathbf{M}(\xi_{n+1}) \neq 0$.

c) The procedure of computing $\xi_0, \xi_1, \ldots$ is stopped at the n-th step, after checking that ξ_n is almost optimal. This is done according to a stopping rule, for example the one given in Proposition IV.28.

An iterative procedure produces a "trajectory" through the points $\mathbf{M}(\xi_n)$; $(n=0, 1, 2, \ldots)$ in the set $\mathfrak{M}_+$ of nonsingular information matrices. Even if the optimum design is singular, its information matrix can be approached by such "regular trajectories". However, difficulties arise in the singular case because of possible discontinuities of the optimality function (see Chapters III.2 and IV.5).

Many methods of computation are known for the case of D-optimality. Two of them are presented in detail in Chapter V.2 and V.3. The one given in Chapter V.2 can also be used for the case of partial D-optimality (particularly also for the case $\phi[\mathbf{M}(\xi)] = \mathrm{var}_\xi g$). The method is a variant of the gradient method. Even though the method is simple to apply, it is difficult to prove its convergence, since the criterion function is not continuous and can achieve the value $+\infty$ in $\mathfrak{M} - \mathfrak{M}_+$. In practical use, even though the computation is simple in each step, it is slow because of the amount of steps required. A faster method is presented in Chapter V.3 but for the case of a global D-optimality on a finite set X. Other methods for D-optimality are in Chapter V.4. In Chapter V.5 general features of iterative methods are discussed for the case of a convex ϕ satisfying the conditions (A), (B) from Chapter IV.4. This general approach is intended as a tool for proving the convergence of different iterative methods. In Chapter V.6 other iterative methods for convex criteria function are given briefly. In Chapter V.7 we outline a method for computing ϕ-optimum designs, subject to the condition that two alternative regression models must be taken into account.

We recall the notation ξ_x for the design supported by one point $x \in X$. Evidently,

$$\mathbf{M}(\xi_x) = \mathbf{f}(x)\mathbf{f}'(x).$$

V.2. COMPUTATION OF A PARTIAL D-OPTIMUM DESIGN

We recall the notation $\phi_0(\mathbf{M}) = \log\det[\mathbf{M}^-]_{\mathrm{I}}$ (if $\alpha_1, \ldots, \alpha_s$ are estimable), $\phi_0(\mathbf{M}) = \infty$ (othervise) introduced first in Eq. (32) in Chapter IV.

The method is based on the following algorithm (cf. [45, 46, 57]):

a) a starting design ξ_0 is chosen; $\det \mathbf{M}(\xi_0) \neq 0$,

b) the iteration at the n-th step is

$$\xi_n = \left(1 - \frac{1}{n}\right) \xi_{n-1} + \frac{1}{n}\, \xi_{x_n}; \quad (x \in X),$$

where $x_n \in X$ is the point where the function

$$x \in X \mapsto \sum_{i,j=1}^{m} f_i(x)\{\mathbf{M}^{-1}(\xi_{n-1})\}_{ij} f_j(x) - \sum_{i,j=s+1}^{m} f_i(x)\{[\mathbf{M}_{\mathrm{III}}(\xi_{n-1})]^{-1}\}_{ij} f_j(x)$$

achieves its maximum.

PROPOSITION V.1. *Let* $\{\xi_n\}$ *be the sequence obtained by the iterative procedure. Then*

$$\lim_{n\to\infty} \phi_0[\mathbf{M}(\xi_n)] = \inf\{\phi_0[\mathbf{M}(\xi)] : \xi \in \Xi\}.$$

Several auxiliary statements (Propositions V.2—V.5) are required for the proof. We present them in detail because they are useful by themselves as well.

We shall use the following notation

$$d(x, \xi) \equiv \mathbf{f}'(x)\mathbf{M}^{-1}(\xi)\mathbf{f}(x),$$

$$d_{s|m}(x, \xi) \equiv \mathbf{f}'(x)\mathbf{M}^{-1}(\xi)\mathbf{f}(x) - \sum_{i,j=s+1}^{m} f_i(x)\{[\mathbf{M}_{\mathrm{III}}(\xi)]^{-1}\}_{ij} f_j(x). \tag{1}$$

Evidently, $d_{m|m}(\cdot) = d(\cdot)$.

PROPOSITION V.2. *Let* $\mathbf{M}(\xi)$ *be nonsingular. Then*

$$s - \partial\phi_0[\mathbf{M}(\xi), \mathbf{f}(x)\mathbf{f}'(x)] = d_{s|m}(x, \xi)$$

for every $x \in X$.

The proof follows from Eq. (60) in Chapter IV and from Proposition IV.14. □

From Proposition V.2 it follows that the passage from $\mathbf{M}(\xi_n)$ to $\mathbf{M}(\xi_{n+1})$ is done in the direction of the steepest descent of the function ϕ_0.

PROPOSITION V.3 [42]. *If* $\det \mathbf{M}(\xi) \neq 0$, *then*

$$0 \leqslant \frac{1}{s}\left(\phi_0[\mathbf{M}(\xi)] - \inf\{\phi_0[\mathbf{M}(\eta)] : \eta \in \Xi\}\right)$$
$$\leqslant \log\left[\max_{x \in X} d_{s|m}(x, \xi)/s\right].$$

PROOF. Let $\mathbf{M} \in \mathfrak{M}_+$. According to Proposition III.12, there exists a nonsingular $s \times s$ matrix $\mathbf{C}$ and a nonsingular $(m-s)\times(m-s)$ matrix $\mathbf{D}$ such that

$$\begin{aligned} &\mathbf{C}(\mathbf{M}_{\mathrm{I}} - \mathbf{M}_{\mathrm{II}}(\mathbf{M}_{\mathrm{III}})^{-1}\mathbf{M}'_{\mathrm{II}})\mathbf{C}' = \mathbf{I}, \\ &\mathbf{D}\mathbf{M}_{\mathrm{III}}\mathbf{D}' = \mathbf{I}. \end{aligned} \tag{2}$$

Define an $m \times m$ matrix $\mathbf{L}$ as

$$\mathbf{L} = \begin{pmatrix} \mathbf{C}, & -\mathbf{C}\mathbf{M}_{\mathrm{II}}(\mathbf{M}_{\mathrm{III}})^{-1} \\ \mathbf{0}, & \mathbf{D} \end{pmatrix}.$$

Using Eq. (2) we can verify that

$$\begin{aligned} &\mathbf{L}\mathbf{M}\mathbf{L}' = \mathbf{I}, \\ &\mathbf{L}^{-1} = \begin{pmatrix} \mathbf{C}^{-1}, & \mathbf{M}_{\mathrm{II}}(\mathbf{M}_{\mathrm{III}})^{-1}\mathbf{D}^{-1} \\ \mathbf{0}, & \mathbf{D}^{-1} \end{pmatrix}. \end{aligned} \tag{3}$$

Hence

$$\mathbf{M}_{\mathrm{III}} = (\mathbf{L}^{-1}\mathbf{L}'^{-1})_{\mathrm{III}} = \mathbf{D}^{-1}\mathbf{D}'^{-1}. \tag{4}$$

We can simplify the proven inequality using Eqs. (3) and (4). For this purpose set a new basis $\varphi_1, \ldots, \varphi_m$ defined as

$$\varphi(x) = \mathbf{L}\boldsymbol{f}(x); \quad (x \in X) \tag{5}$$

instead of the basis $f_1, \ldots, f_m$. The information matrix $\mathbf{M} \equiv \mathbf{M}(\xi)$ is transformed in this way into the identity matrix

$$\mathbf{M}_\varphi(\xi) \equiv \sum_{x \in X} \varphi(x)\varphi'(x)\xi(x) = \mathbf{L}\mathbf{M}\mathbf{L}' = \mathbf{I}.$$

The expression for $d_{s|m}(x, \xi)$, as well as the difference $\phi_0[\mathbf{M}(\xi)] -$

$\phi_0[\mathbf{M}(\eta)]$, are invariant to this change of the basis, which can be verified setting $\mathbf{f}(x)$ from Eq. (5) into Eq. (1) and into Eq. (46), Chapter IV, and using Eq. (46) to express $\phi_0[\mathbf{M}(\xi)]$ and $\phi_0[\mathbf{M}(\eta)]$.

Thus it can be supposed without lack of generality that $\mathbf{M}(\xi)=\mathbf{I}$. Let η be another design, $\det \mathbf{M}(\eta)\neq 0$. Then

$$\begin{aligned}\max_{x\in X} d_{s|m}(x,\xi) &\geq \sum_{x\in X} d_{s|m}(x,\xi)\eta(x)\\ &= \operatorname{tr} \mathbf{M}(\eta) - \operatorname{tr} \mathbf{M}_{\mathrm{III}}(\eta)\\ &= \operatorname{tr} \mathbf{M}_{\mathrm{I}}(\eta). \end{aligned} \tag{6}$$

Let $\lambda_1, \ldots, \lambda_s$ be the eigenvalues of $[\mathbf{M}^{-1}(\eta)]_{\mathrm{I}}$. We can write

$$\begin{aligned}\frac{1}{s}(\phi_0[\mathbf{M}(\xi)]-\phi_0[\mathbf{M}(\eta)]) &= \frac{1}{s}(\log\det \mathbf{I} - \log\det [\mathbf{M}^{-1}(\eta)]_{\mathrm{I}}) =\\ &= \frac{1}{s}\log \prod_{i=1}^{s} \lambda_i^{-1} \leq \log\left(\frac{1}{s}\sum_{i=1}^{s}\lambda_i^{-1}\right)\\ &= \log\left(\frac{1}{s}\operatorname{tr}\{[\mathbf{M}^{-1}(\eta)]_{\mathrm{I}}\}^{-1}\right), \end{aligned} \tag{7}$$

after using the inequality between the arithmetic and the geometric means. From Eq. (45), Chapter IV it follows that

$$\begin{aligned}&([\mathbf{M}^{-1}(\eta)]_{\mathrm{I}})^{-1}\\ &= \mathbf{M}_{\mathrm{I}}(\eta) - \mathbf{M}_{\mathrm{II}}(\eta)[\mathbf{M}_{\mathrm{III}}(\eta)]^{-1}\mathbf{M}_{\mathrm{II}}'(\eta).\end{aligned}$$

Hence

$$[\operatorname{tr}(\mathbf{M}^{-1}(\eta))_{\mathrm{I}}]^{-1} \leq \operatorname{tr} \mathbf{M}_{\mathrm{I}}(\eta)$$

which allows to compare Eq. (6) with Eq. (7). Thus

$$\begin{aligned}&\frac{1}{s}\{\phi_0[\mathbf{M}(\xi)]-\phi_0[\mathbf{M}(\eta)]\}\\ &\leq \log\left\{\frac{1}{s}\max_{x\in X} d_{s|m}(x,\xi)\right\}; \quad (\mathbf{M}(\eta)\in\mathfrak{M}_+).\end{aligned}$$

We shall end the proof by verifying the equality

$$\inf\{\phi_0(\mathbf{M}): \mathbf{M}\in\mathfrak{M}\} = \inf\{\phi_0(\mathbf{M}): \mathbf{M}\in\mathfrak{M}_+\}.$$

For that take $\mathbf{M}_1 \in \mathfrak{M} - \mathfrak{M}_+$ such that $\phi_0(\mathbf{M}_1) < \infty$. The matrix

$$\mathbf{M}_\varepsilon \equiv (1-\varepsilon)\mathbf{M}_1 + \varepsilon \mathbf{M}(\xi)$$
$$= (1-\varepsilon)\mathbf{M}_1 + \varepsilon \mathbf{I}$$

is nonsingular for every $\varepsilon > 0$. The function

$$\psi\colon \varepsilon \in \langle 0, 1\rangle \mapsto \phi_0(\mathbf{M}_\varepsilon)$$

is convex (Proposition IV.14), hence

$$\phi_0(\mathbf{M}_1) = \psi(0) \geqslant \lim_{\varepsilon \searrow 0} \psi(\varepsilon) \geqslant \inf \{\phi_0(\mathbf{M}) : \mathbf{M} \in \mathfrak{M}_+\}.$$

It follows that

$$\inf \{\phi_0(\mathbf{M}) : \mathbf{M} \in \mathfrak{M}\} \geqslant \inf \{\phi_0(\mathbf{M}) : \mathbf{M} \in \mathfrak{M}_+\}.$$

The inverse inequality is evident. □

EXERCISE V.1. Based on the value of $\max_{x \in X} d_{s|m}(x, \xi)$ and using Proposition V.3 it can be checked how far is the checked design ξ from a D-optimum design. Compare this result with Proposition IV.28. Show that for the case of D-optimality

$$s \log \{\max_x d_{s|m}(x, \xi)/s\} \leqslant -\min_x \partial\phi[\mathbf{M}(\xi), \mathbf{f}(x)\mathbf{f}'(x)],$$

which implies that the statement in Proposition V.3 is stronger than that in Proposition IV.28.

HINT. Use Proposition V.2, the expression (60) for the directional derivative, and look for the minimum of the function $x \in (-\infty, \infty) \mapsto x - 1 - \log x$.

The important expression $d_{s|m}(x, \xi)$ contains the inverse matrices $\mathbf{M}^{-1}(\xi)$, $[\mathbf{M}_{\mathrm{III}}(\xi)]^{-1}$. Programs inverting matrices take a lot of computing time. For that reason the following proposition is useful.

PROPOSITION V.4 [18]. *Let* $\eta = (1-\beta)\xi + \beta\xi_x$, *where* $\beta \in (0, 1)$, $\xi \in \Xi$. *Suppose that* $\det \mathbf{M}(\xi) \neq 0$. *Then* $\det \mathbf{M}(\eta) \neq 0$, *and*

$$\mathbf{M}^{-1}(\eta) = \frac{1}{1-\beta}\left\{\mathbf{M}^{-1}(\xi) - \frac{\beta \mathbf{M}^{-1}(\xi)\mathbf{f}(x)\mathbf{f}'(x)\mathbf{M}^{-1}(\xi)}{1-\beta+\beta d(x, \xi)}\right\}, \quad (8)$$

$$[\mathbf{M}_{\mathrm{III}}(\eta)]^{-1}=\frac{1}{1-\beta}\left\{[\mathbf{M}_{\mathrm{III}}(\xi)]^{-1}\right.$$
$$\left.-\frac{\beta[\mathbf{M}_{\mathrm{III}}(\xi)]^{-1}\boldsymbol{f}^{(2)}(x)\boldsymbol{f}^{(2)\prime}(x)[\mathbf{M}_{\mathrm{III}}(\xi)]^{-1}}{1-\beta+\beta\boldsymbol{f}^{(2)\prime}(x)[\mathbf{M}_{\mathrm{III}}(\xi)]^{-1}\boldsymbol{f}^{(2)}(x)}\right\}, \tag{9}$$

where $\boldsymbol{f}^{(2)}(x)=(f_{s+1}(x), \ldots, f_m(x))'$. *Furthermore,*

$$\det[\mathbf{M}^{-1}(\eta)]_{\mathrm{I}}$$
$$=\det[\mathbf{M}^{-1}(\xi)]_{\mathrm{I}}(1-\beta)^{-s}\left[1-\frac{\beta d_{s|m}(x,\xi)}{1-\beta+\beta d(x,\xi)}\right]. \tag{10}$$

PROOF. The matrix $\mathbf{M}(\xi)$ is nonsingular hence positive definite. It follows that the matrix

$$\mathbf{M}(\eta)=(1-\beta)\mathbf{M}(\xi)+\beta\boldsymbol{f}(x)\boldsymbol{f}'(x)$$

is positive definite hence nonsingular as well. Thus

$$\mathbf{M}^{-1}(\eta)$$
$$=\frac{\mathbf{M}^{-1}(\xi)}{1-\beta}\left[\mathbf{I}+\frac{\beta}{1-\beta}\boldsymbol{f}(x)\boldsymbol{f}'(x)\mathbf{M}^{-1}(\xi)\right]^{-1}. \tag{11}$$

Using Eq. (13) in Chapter III, we obtain after some rearrangements from Eq. (11) the required Eq. (8).

Similarly, from

$$\mathbf{M}_{\mathrm{III}}(\eta)=(1-\beta)\mathbf{M}_{\mathrm{III}}(\xi)+\beta\boldsymbol{f}^{(2)}(x)\boldsymbol{f}^{(2)\prime}(x)$$

the expression in Eq. (9) is obtained.

Consider the matrix

$$\mathbf{H}=\begin{pmatrix}\mathbf{A}_1, & \mathbf{B}_1\\ \mathbf{C}_1, & \mathbf{D}_1\end{pmatrix},$$

where $\mathbf{A}_1$ is a nonsingular $p\times p$ matrix and $\mathbf{D}_1$ is a $q\times q$ matrix. The determinant of $\mathbf{H}$ is unaffected when multiplying the block-row $(\mathbf{A}_1, \mathbf{B}_1)$ by the matrix $\mathbf{C}_1\mathbf{A}_1^{-1}$ and when subtracting it from the block-row $(\mathbf{C}_1, \mathbf{D}_1)$. We obtain

$$\det\mathbf{H}=\det\begin{pmatrix}\mathbf{A}_1, & \mathbf{B}_1\\ \mathbf{0}, & \mathbf{D}_1-\mathbf{C}_1\mathbf{A}_1^{-1}\mathbf{B}_1\end{pmatrix}=\det\mathbf{A}_1\det(\mathbf{D}_1-\mathbf{C}_1\mathbf{A}_1^{-1}\mathbf{B}_1). \tag{12}$$

Using Eq. (12) twice, we obtain

$$(1-\beta)^m \det \mathbf{M}(\xi) \det\left[1+\frac{\beta}{1-\beta}\, \mathbf{f}'(x)\mathbf{M}^{-1}(\xi)\mathbf{f}(x)\right]$$

$$=\det\begin{pmatrix}(1-\beta)\mathbf{M}(\xi), & \sqrt{\beta}\mathbf{f}(x)\\ -\sqrt{\beta}\mathbf{f}'(x), & 1\end{pmatrix}$$

$$=\det\begin{pmatrix}1, & -\sqrt{\beta}\mathbf{f}'(x)\\ \sqrt{\beta}\mathbf{f}(x), & (1-\beta)\mathbf{M}(\xi)\end{pmatrix}$$

$$=1\ \ \det[(1-\beta)\mathbf{M}(\xi)+\beta\mathbf{f}(x)\mathbf{f}'(x)]=\det\mathbf{M}(\eta).$$

Similarly,

$$(1-\beta)^{m-s}\det \mathbf{M}_{\mathrm{III}}(\xi)$$
$$\times \det\left(1+\frac{\beta}{1-\beta}\,\mathbf{f}^{(2)\prime}(x)[\mathbf{M}_{\mathrm{III}}(\xi)]^{-1}\mathbf{f}^{(2)}(x)\right)=\det\mathbf{M}_{\mathrm{III}}(\eta).$$

Applying Eq. (46) from Chapter IV, we obtain

$$\det[\mathbf{M}^{-1}(\eta)]_{\mathrm{I}}=\frac{\det\mathbf{M}_{\mathrm{III}}(\eta)}{\det\mathbf{M}(\eta)}$$

$$=(1-\beta)^{-s}\det[\mathbf{M}^{-1}(\xi)]_{\mathrm{I}}\left[1-\frac{\beta d_{s|m}(x,\xi)}{1-\beta+\beta d(x,\xi)}\right].\ \square$$

PROPOSITION V.5 [46]. *Let $\{x_n\}$ be an arbitrary sequence of points from X, such that the matrix*

$$\mathbf{M}_n=\sum_{i=1}^{n}\mathbf{f}(x_i)\mathbf{f}'(x_i)$$

is nonsingular at least for one n (and consequently for every $n'>n$). Then there is the limit

$$\lim_{n\to\infty}\mathbf{f}'(x_n)\mathbf{M}_n^{-1}\mathbf{f}(x_n)=0. \tag{13}$$

PROOF [62]. Let N_1 be an arbitrary infinite sequence of positive integers. Since $\mathbf{f}(X)$ is a compact set, we can write

$$\lim_{\substack{n\to\infty\\ n\in N_2}}\mathbf{f}(x_n)=\mathbf{f}(\bar{x}) \tag{14}$$

for some $N_2 \subset N_1$ and for a point $\bar{x} \in X$. According to Eq. (7) in Chapter II,

$$\mathbf{f}'(\bar{x})\mathbf{M}_n^{-1}\mathbf{f}(\bar{x}) = \max\left\{\frac{[\mathbf{u}'\mathbf{f}(\bar{x})]^2}{\mathbf{u}'\mathbf{M}_n\mathbf{u}} : \mathbf{u} \in R^m,\ \|\mathbf{u}\| = 1\right\}. \tag{15}$$

Define the functions $\psi_1, \psi_2, \ldots$

$$\psi_n(\mathbf{u}) \equiv [\mathbf{f}'(\bar{x})\mathbf{u}]^2/(\mathbf{u}'\mathbf{M}_n\mathbf{u}); \quad (n = 1, 2, \ldots)$$

on the compact set

$$\{\mathbf{u}: \mathbf{u} \in R^m,\ \|\mathbf{u}\| = 1\}.$$

These functions are continuous, the sequence $\{\psi_n(\mathbf{u})\}_{n=1}^{\infty}$ is nonincreasing, and $\lim_{n\to\infty} \psi_n(\mathbf{u}) = 0$. From the Dinni theorem of functional analysis (cf. [112]) it follows that this convergence is uniform, hence

$$\lim_{n\to\infty} \max\{\psi_n(\mathbf{u}) : \mathbf{u} \in R^m,\ \|\mathbf{u}\| = 1\} = 0.$$

This, together with Eq. (15), implies

$$\lim_{n\to\infty} \mathbf{f}'(\bar{x})\mathbf{M}_n^{-1}\mathbf{f}(\bar{x}) = 0$$

from which it follows, using Eq. (14), that

$$\begin{aligned} 0 &\leq \limsup_{\substack{n\to\infty \\ n\in N_2}} \mathbf{f}'(x_n)\mathbf{M}_n^{-1}\mathbf{f}(x_n) \\ &\leq \lim_{n\to\infty} \mathbf{f}'(\bar{x})\mathbf{M}_n^{-1}\mathbf{f}(\bar{x}) + \limsup_{\substack{n\to\infty \\ n\in N_2}} |\mathbf{f}'(x_n)\mathbf{M}_n^{-1}\mathbf{f}(x_n) \\ &\quad - \mathbf{f}'(\bar{x})\mathbf{M}_n^{-1}\mathbf{f}(\bar{x})| = 0, \end{aligned}$$

since the sequence of norms $\{\|\mathbf{M}_n^{-1}\|\}_n$ is nonincreasing. Summing up: each subsequence $\{x_n\}_{n\in N_1}$ of the original sequence $\{x_n\}_{n=1}^{\infty}$ has by itself a subsequence $\{x_n\}_{n\in N_2}$ such that

$$\lim_{n\in N_2} \mathbf{f}'(x_n)\mathbf{M}_n^{-1}\mathbf{f}(x_n) = 0,$$

which implies the desired convergence in Eq. (13). □

PROOF OF PROPOSITION V.1. Set $\xi=\xi_n$, $\eta=\xi_{n+1}$, $\beta=(n+1)^{-1}$, $x=x_{n+1}$ into Eq. (10). We obtain

$$\frac{\det[\mathbf{M}^{-1}(\xi_{n+1})]_{\mathrm{I}}}{\det[\mathbf{M}^{-1}(\xi_n)]_{\mathrm{I}}}=\left(\frac{n+1}{n}\right)^s\left[1-\frac{1}{n}\frac{d_{s|m}(x_{n+1},\xi_n)}{1+\frac{1}{n}d(x_{n+1},\xi_n)}\right].$$

According to Proposition V.3

$$d_{s|m}(x_{n+1},\xi_n)\geqslant s\left[\frac{\det[\mathbf{M}^{-1}(\xi_n)]_{\mathrm{I}}}{\inf_{\xi}\det[\mathbf{M}^{-1}(\xi)]_{\mathrm{I}}}\right]^{1/s},$$

hence

$$\frac{\det[\mathbf{M}^{-1}(\xi_{n+1})]_{\mathrm{I}}}{\det[\mathbf{M}^{-1}(\xi_n)]_{\mathrm{I}}}\leqslant\left(\frac{n+1}{n}\right)^s\left[1-\frac{s}{n+d(x_{n+1},\xi_n)}\left(\frac{\det[\mathbf{M}^{-1}(\xi_n)]_{\mathrm{I}}}{\inf_{\xi}\det[\mathbf{M}^{-1}(\xi)]_{\mathrm{I}}}\right)^{1/s}\right]. \tag{16}$$

Choose an arbitrary $\varepsilon>0$. We have

$$\lim_{n\to\infty}\left(\frac{n+1}{n}\right)^s\left(1-\frac{s}{n}\frac{1}{(1+\varepsilon/4)^{1/s}}\right)=1,$$

$$\lim_{n\to\infty}d(x_{n+1},\xi_n)/n=0$$

(Proposition V.5). Hence there is n_0 such that for $n\geqslant n_0$

$$\left(\frac{n+1}{n}\right)^s\left(1-\frac{s}{n}\frac{1}{(1+\varepsilon/4)^{1/s}}\right)<1+\varepsilon/3, \tag{17}$$

$$1+n^{-1}d(x_{n+1},\xi_n)<(1+\varepsilon/4)^{1/s}. \tag{18}$$

Let $q\equiv[(1+\varepsilon/3)/(1+\varepsilon/4)]^{1/s}>1$. Expanding $(1+1/n)^s$ according to the binomial formula, we obtain

$$\left(1+\frac{1}{n}\right)^s\left(1-\frac{sq}{n}\right)=1+\sum_{i=1}^{s}\left[\binom{s}{i}-sq\binom{s}{i-1}\right]\left(\frac{1}{n}\right)^i-sq\left(\frac{1}{n}\right)^{s+1}<1+\frac{s}{n}(1-q). \tag{19}$$

Suppose that

$$\frac{\det [\mathbf{M}^{-1}(\xi_n)]_{\mathrm{I}}}{\inf_{\xi} \det [\mathbf{M}^{-1}(\xi)]_{\mathrm{I}}} \geqslant 1+\varepsilon/3 \tag{20}$$

for some $n > n_0$. Then, according to Eqs. (16) and (19),

$$\frac{\det [\mathbf{M}^{-1}(\xi_{n+1})]_{\mathrm{I}}}{\det [\mathbf{M}^{-1}(\xi_n)]_{\mathrm{I}}} \leqslant \left(\frac{n+1}{n}\right)^s \left[1-\frac{s}{n}\left(\frac{(1+\varepsilon/3)^{1/s}}{(1+\varepsilon/4)^{1/s}}\right)\right] < 1+\frac{s}{n}(1-q) < 1.$$

This inequality guarantees that

$$1 \leqslant \frac{\det [\mathbf{M}^{-1}(\xi_k)]_{\mathrm{I}}}{\inf_{\xi} \det [\mathbf{M}^{-1}(\xi)]_{\mathrm{I}}} < 1+\varepsilon/3$$

for some $k > n$. From Eqs. (16)—(18) it follows that

$$\frac{\det [\mathbf{M}^{-1}(\xi_{k+1})]_{\mathrm{I}}}{\det [\mathbf{M}^{-1}(\xi_k)]_{\mathrm{I}}} \leqslant \left(\frac{k+1}{k}\right)^s \left[1-\frac{s}{k}\frac{1}{(1+\varepsilon/4)^{1/s}}\right] < 1+\varepsilon/3,$$

hence

$$\frac{\det [\mathbf{M}^{-1}(\xi_{k+1})]_{\mathrm{I}}}{\inf_{\xi} \det [\mathbf{M}^{-1}(\xi)]_{\mathrm{I}}} < (1+\varepsilon/3)^2 < 1+\varepsilon.$$

Summing up: to every $\varepsilon > 0$ there is an n_ε such that for all $n \geqslant n_\varepsilon$

$$\phi_0[\mathbf{M}(\xi_n)] - \inf_{\xi} \phi_0[\mathbf{M}(\xi)] < \log(1+\varepsilon). \ \square$$

V.3. THE COMPUTATION OF A D-OPTIMUM DESIGN ON A FINITE SET X

The rate of convergence of the method from Chapter V.2 is rather low, hence other methods have been proposed for particular cases. In case of global D-optimality on a finite set

$$X=\{x_1, \ldots, x_k\}$$

a method can be based on the mapping $T: \Xi \mapsto \Xi$ defined by

$$(T\xi)(x_i)=\xi(x_i)\frac{d(x_i, \xi)}{m}; \quad (i=1, \ldots, k).$$

From Proposition IV.6 it follows that a design ξ^* is D-optimal exactly if it is a fixed point of the mapping T, i.e.

$$T\xi^*=\xi^*.$$

Therefore, in [47] the following algorithm has been proposed:

a) We start by any design ξ_0 such that $\xi_0(x_i)>0$; $(i=1, \ldots, k)$.

b) At the n-th step

$$\xi_{n+1}(x_i)=\xi_n(x_i)\frac{d(x_i, \xi_n)}{m}; \quad (i=1, \ldots, k).$$

c) Following any of the stopping rules in Proposition V.3 or Proposition IV.28, the computation is stopped when

$$\frac{d(x_i, \xi_n)}{m}<1+\varepsilon; \quad (i=1, \ldots, k).$$

PROPOSITION V.6. *If* $\{\xi_n\}$ *is a sequence of designs obtained according to* a) *and* b), *then the sequence* $\{\det \mathbf{M}(\xi_n)\}_{n=0}^{\infty}$ *is nondecreasing, and it converges to* $\max\{\det \mathbf{M}(\xi): \xi\in\Xi\}$.

The proof requires several auxiliary statements.

PROPOSITION V.7 [48]. *If* $\xi\in\Xi$, $\det \mathbf{M}(\xi)\neq 0$, *then*

$$\xi(x)d(x, \xi)\leq 1; \quad (x\in X).$$

PROOF. When $\xi(x)=0$, the inequality is evident. In case of $\xi(x)>0$, choose $\varepsilon\in(0, \xi(x))$, and denote $\beta=\xi(x)-\varepsilon$. Define

$$\xi_0(x')=\begin{cases} \varepsilon/(1-\beta) \text{ if } x'=x, \\ \xi(x')/(1-\beta) \text{ if } x'\neq x. \end{cases}$$

Obviously, $\det \mathbf{M}(\xi_0) \neq 0$ and

$$\xi = (1-\beta)\xi_0 + \beta\xi_x.$$

Set into Eq. (8) ξ_0 instead of ξ and ξ instead of η. Multiply the obtained inequality by $\mathbf{f}'(x)$ from the left and by $\mathbf{f}(x)$ from the right. We obtain

$$d(x, \xi) = \frac{d(x, \xi_0)}{1-\beta+\beta d(x, \xi_0)},$$

hence

$$(\xi(x)-\varepsilon)d(x, \xi) = \beta d(x, \xi) < 1.$$

With ε tending to zero, we obtain the desired inequality. □

Proposition V.8. *If* $\det \mathbf{M}(\xi) \neq 0$, *then*

$$\det \mathbf{M}(\xi) = \sum_{i_1=1}^{k} \cdots \sum_{i_m=1}^{k} h(x_{i_1}, \ldots, x_{i_m})\xi(x_{i_1}) \ldots \xi(x_{i_m}),$$

where

$$h(x_1, \ldots, x_m) = \frac{1}{m!} \det{}^2[\mathbf{f}(x_1), \ldots, \mathbf{f}(x_m)].$$

Proof. Obviously,

$$\det(\{\mathbf{A}\}_{\cdot 1}, \ldots, \{\mathbf{A}\}_{\cdot i-1}, c\{\mathbf{A}\}_{\cdot i}, \{\mathbf{A}\}_{\cdot i+1}, \ldots, \{\mathbf{A}\}_{\cdot m}) = c \det \mathbf{A}$$

for any $m \times m$ matrix $\mathbf{A}$ and any $c \in R$.
Hence

$$\begin{aligned}\det \mathbf{M}(\xi) &= \det\left\{\sum_{i_1=1}^{k} \mathbf{f}(x_{i_1})\xi(x_{i_1})f_1(x_{i_1}), \ldots, \sum_{i_m=1}^{k} \mathbf{f}(x_{i_m})\xi(x_{i_m})f_m(x_{i_m})\right\} \\ &= \sum_{i_1=1}^{k} \cdots \sum_{i_1=1}^{k} \xi(x_{i_1}) \ldots \xi(x_{i_m}) \\ &\times f_1(x_{i_1}) \ldots f_m(x_{i_m}) \det(\mathbf{f}(x_{i_1}), \ldots, \mathbf{f}(x_{i_m})).\end{aligned} \tag{21}$$

Summing up terms in (21) which have the same subscripts $i_1, \ldots, i_m$ but in different orderings, we obtain the desired result. □

PROPOSITION V.9. *Let $w_1, \ldots, w_m$ be independent random variables which have values in X and are distributed according to ξ. Let h: $(w_1, \ldots, w_m) \in X^m \mapsto R$ be a nonnegative function. Then*

$$[\mathrm{E}(h)]^{m+1} \leq \mathrm{E}\{h\,\mathrm{E}(h|w_1) \ldots \mathrm{E}(h|w_m)\},$$

and the equality is strict unless

$$\mathrm{E}(h|w_1) = \ldots = \mathrm{E}(h|w_m) = \mathrm{E}(h)$$

almost everywhere $[\xi]$.

PROOF. From the inequality $t-1 \geq \log t$ we obtain

$$\begin{aligned} &\prod_{i=1}^{m} [\mathrm{E}(h|w_i)/\mathrm{E}(h)] - 1 \\ &\geq \sum_{i=1}^{m} [\log \mathrm{E}(h|w_i) - \log \mathrm{E}(h)]. \end{aligned} \tag{22}$$

Multiply Eq. (22) by h. We obtain

$$\begin{aligned} &\frac{1}{[\mathrm{E}(h)]^m} \mathrm{E}\left\{h \prod_{i=1}^{m} \mathrm{E}(h|w_i)\right\} - \mathrm{E}(h) \\ &\geq \sum_{i=1}^{m} \{\mathrm{E}[h \log \mathrm{E}(h|w_i)] - \mathrm{E}(h) \log \mathrm{E}(h)\} \\ &= \sum_{i=1}^{m} \{\mathrm{E}[\mathrm{E}(h|w_i) \log \mathrm{E}(h|w_i)] - \mathrm{E}(h) \log \mathrm{E}(h)\}. \end{aligned}$$

The function $t \in (0, \infty) \mapsto \varphi(t) \equiv t \log t$ is strictly convex. Using the Jensen inequality

$$\mathrm{E}[\varphi(t)] \geq \varphi[\mathrm{E}(t)],$$

we obtain that

$$\mathrm{E}[\mathrm{E}(h|w_i) \log \mathrm{E}(h|w_i)] \geq \mathrm{E}(h) \log \mathrm{E}(h)$$

with strict inequality unless $\mathrm{E}(h|w_i) = \mathrm{E}(h)$ for almost every w_i. □

PROOF OF PROPOSITION V.6 ([47]). From Propositions IV.2 and V.8 it follows that

$$\frac{d(x_i, \xi)}{m} = \frac{\partial \ln \det \mathbf{M}(\xi)}{\partial \xi(x_i)} \Bigg/ \left[\sum_{j=1}^{k} \xi(x_j) \frac{\partial \ln \det \mathbf{M}(\xi)}{\partial \xi(x_j)}\right]$$
$$= \frac{\partial \, \mathrm{E}_\xi\{h\}}{\partial \xi(x_i)} \Bigg/ \sum_{j=1}^{k} \xi(x_j) \frac{\partial \, \mathrm{E}_\xi\{h\}}{\partial \xi(x_j)},$$

when taking h as in Proposition V.8. Obviously,

$$[\partial \, \mathrm{E}_\xi(h)/\partial \xi(x_i)] = \mathrm{E}_\xi(h|x_i).$$

Thus

$$\frac{d(x_i, \xi)}{m} = \frac{\mathrm{E}_\xi(h|x_i)}{\mathrm{E}_\xi(h)}.$$

Hence, applying Proposition V.9, we obtain

$$\det \mathbf{M}(T\xi) = \mathrm{E}_{T\xi}(h) = \frac{\mathrm{E}_\xi\left[h \prod_{i=1}^{m} \mathrm{E}_\xi(h|x_i)\right]}{[\mathrm{E}_\xi(h)]^m}$$
$$\geqslant \mathrm{E}_\xi(h) = \det \mathbf{M}(\xi) \tag{23}$$

with equality exactly if $T\xi = \xi$. A consequence of Eq. (23) is that the sequence $\{\det \mathbf{M}(\xi_n)\}_n$ is nondecreasing.

The sequence $\{\xi_n\}_n$ is bounded, since $0 \leqslant \xi_n(x) \leqslant 1$; $(x \in X)$. Hence there is at least one limit point of $\{\xi_n\}_n$. Furthermore, all limit points of $\{\xi_n\}_n$ have the determinant of the information matrix equal to

$$\lim_{n \to \infty} \det \mathbf{M}(\xi_n).$$

Let $\{\xi_n\}_{n \in N_1}$ be a subsequence converging to ξ^*, and let, in turn, $\{\xi_n\}_{n \in N_2}$ be a subsequence of $\{\xi_n\}_{n+1 \in N_1}$ converging to some ξ. From the equality

$$\xi_n(x) = \xi_{n-1}(x) \frac{d(x, \xi_{n-1})}{m}$$

it follows that

$$\xi^*(x) = \xi(x) \frac{d(x, \xi)}{m} = (T\xi)(x).$$

Since det $\mathbf{M}(\xi^*)=\det \mathbf{M}(\xi)$, Eq. (23) implies $\xi^*=T\xi=\xi$, hence ξ^* is D-optimal. Thus

$$\lim_{n\to\infty} \det \mathbf{M}(\xi_n)=\max\{\det \mathbf{M}(\xi):\ \xi\in\Xi\}.\ \square$$

V.4. OTHER ITERATIVE METHODS FOR D-OPTIMALITY

In this chapter other methods are presented briefly without proofs.

V.4.1. *The Steepest Descent Method for Global D-optimality*

The description of the algorithm is (cf. [6], Chapter 2.5):

a) Start with ξ_0, $\det \mathbf{M}(\xi_0)\neq 0$.

b) At the n-th step compute

$$\xi_{n+1}=(1-\beta_n)\xi_n+\beta_n\xi_{x_n},$$

where x_n and β_n are taken so that the maxima of the functions

$$x\in X\mapsto \boldsymbol{f}'(x)\mathbf{M}^{-1}(\xi_n)\boldsymbol{f}(x)\equiv d(x,\xi_n)$$

and

$$\beta\in\langle 0,1\rangle\mapsto \det \mathbf{M}[(1-\beta)\xi_n+\beta\xi_{x_n}] \tag{24}$$

are achieved at the points x_n and β_n, respectively.

c) Stop computing according to the rules in Propositions IV.28 or V.3. In [6], Chapter 2.5 it has been proven that $\{\det \mathbf{M}(\xi_n)\}_{n=0}^{\infty}$ converges monotonically to $\max\{\det \mathbf{M}(\xi):\ \xi\in\Xi\}$.

Remarks to the algorithm:

The computation of $\mathbf{M}^{-1}(\xi_{n+1})$ can be simplified using Proposition V.4. Furthermore, the number β_n can be computed according to the formula

$$\beta_n=\frac{d(x_n,\xi_n)-m}{m[d(x_n,\xi_n)-1]}. \tag{25}$$

The rate of convergence is comparable with the rate of the method in Chapter V.2.

In case of partial D-optimality a similar method is described in [6], but for the rare case of the partial D-optimum design being nonsingular.

V.4.2. *The Accelerated Method of Steepest Descent*

In methods described in Section 1 or Chapter V.2, the design ξ_n is perturbed by adding the measure β_n (or $1/n$) at the point x_n and by subtracting this measure uniformly from the other points of the support of ξ_n. By this approach still more points are added to the support of the design, and the design measure at noninformative points is decreasing very slowly. The iterative computation can be accelerated when subtracting directly the design measure from noninformative points. This is the basic idea of the following algorithm [48]:

a) Start with ξ_0, det $\mathbf{M}(\xi_0) \neq 0$.

b) Define

$$\xi_{n+1} = (1-\beta_n)\xi_n + \beta_n \xi_{x_n}, \tag{26}$$

where x_n, β_n are

(i) either the points of the maxima of the functions

$$x \in X \mapsto d(x, \xi_n) \tag{27}$$

and

$$\beta \in \langle 0, 1\rangle \mapsto \det [\mathbf{M}(1-\beta)\xi_n + \beta \xi_{x_n}];$$

(ii) or x_n is the point of the minimum of

$$x \in X_{\xi_n} \mapsto d(x, \xi_n), \tag{28}$$

and the choice of β_n is split again, depending on the value of $d(x_n, \xi_n)$. If

$$d(x_n, \xi_n) \geq m/(1+(m-1)\xi_n(x_n)),$$

then β_n is the point of the maximum of

$$\beta \in \left\langle \frac{-\xi_n(x_n)}{1-\xi_n(x_n)}, 1 \right\rangle \mapsto \det \mathbf{M}[(1-\beta)\xi_n + \beta \xi_{x_n}]. \tag{29}$$

Otherwise

$$\beta_n = -\xi_n(x_n).$$

If det $\mathbf{M}[(1-\beta_n)\xi_n + \beta_n \xi_{x_n}]$ is larger for the method (i) than for (ii), compute ξ_{n+1} according to (i); otherwise compute ξ_{n+1} according to (ii).

c) The computation is stopped according to the rules in Propositions IV.28 or V.3.

In [48] it has been proven that the sequence det $\{\mathbf{M}(\xi_n)\}_{n=1}^{\infty}$ is nondecreasing and that it converges to

$$\max \{\det \mathbf{M}(\xi) : \xi \in \Xi\}.$$

REMARKS. It is recommended to use the formula

$$\xi_{n+1} = (1-\beta_n)\xi_n + \beta_n \varkappa_n$$

instead of Eq. (26). Here, $\varkappa_n$ is a design spread uniformly at the points in which the maximum of $d(x, \xi_n)$ is achieved (in the case (i)) or at points in which the minimum of $d(x, \xi_n)$ is achieved in X_{ξ_n} (in the case (ii)). In the case (i), β_n is given by Eq. (25). In the case (ii), the number $\gamma_n \equiv \beta_n(1-\beta_n)$ is given by

$$\gamma_n = \frac{d(x_n, \xi_n) - m}{[d(x_n, \xi_n)(m-1)]^{m-1}}.$$

The computation is simplified when using the equality

$$\det \mathbf{M}[(1-\beta_n)\xi_n + \beta_n \xi_{x_n}] = \frac{\det \mathbf{M}(\xi_n)[1+\gamma_n d(x_n, \xi_n)]}{(1+\gamma_n)^m}.$$

Note that a modified method can also be used in case of partial D-optimality, as stated in [48].

V.4.3. *The Application of the Duality Principle*

The principle of duality known from convex programming has been applied to D-optimality in [49]. A geometrical approach to iterative computation of D-optimum designs which is based on duality is described in [51]. To outline this approach consider the ellipsoid

$$\mathcal{O}_\xi \equiv \{\mathbf{v}: \mathbf{v} \in R^m, \mathbf{v}'\mathbf{M}^{-1}(\xi)\mathbf{v} \leq m\}$$

which contains in particular those vectors $\mathbf{f}(x)$; $(x \in X)$ that satisfy the inequality

$$\mathbf{f}'(x)\mathbf{M}^{-1}(\xi)\mathbf{f}(x) \equiv d(x, \xi) \leq m.$$

The ellipsoid $\mathcal{O}_\xi$ has common points with the set $\mathbf{f}(X)$, since

$$\sum_{x \in X} d(x, \xi)\xi(x) = m.$$

According to the equivalence theorem (Proposition IV.6), $\mathbf{f}(X)$ is a subset of $\mathcal{O}_\xi$ exactly if ξ^* is D-optimal. Then points of $\mathbf{f}(X)$ on the boundary of the ellipsoid $\mathcal{O}_{\xi^*}$ are points supporting D-optimum designs.

On the other hand, the volume of $\mathcal{O}_\xi$ is proportional to $[\det \mathbf{M}(\xi)]^{1/2}$ (compare with Chapter IV.2). Hence, to optimize the design means to maximize the volume of $\mathcal{O}_\xi$. A dual problem is to minimize the volume of an ellipsoid

$$\mathcal{O}_\mathbf{W} \equiv \{\mathbf{v}: \mathbf{v} \in R^m, \mathbf{v}'\mathbf{W}\mathbf{v} \leq m\} \tag{30}$$

subject to the restriction that $\mathbf{f}(X) \subset \mathcal{O}_\mathbf{W}$.

In [51] a procedure is proposed based on constructing a minimum-volume ellipsoid $\mathcal{O}_\mathbf{W}$ subject to the restriction that $\{\mathbf{f}(x): \xi(x) > 0\} \subset \mathcal{O}_\mathbf{W}$. If $\mathbf{f}(X) \subset \mathcal{O}_\mathbf{W}$, then ξ is D-optimal.

V.5. THE COMPUTATION OF ϕ-OPTIMUM DESIGNS FOR CONVEX GLOBAL CRITERIA FUNCTIONS

Iterative methods can be studied geometrically, considering the trajectories in $\mathfrak{M}$ generated by iterative procedures. For general features of such geometrical approach in convex minimization cf. e.g. [110].

We shall suppose in this chapter that the considered criterion function ϕ has the properties (A), (B) from Chapter IV.4.

We recall that $\mathscr{L}(\mathfrak{M})$ is the minimum linear space of symmetric $m \times m$ matrices containing the set $\mathfrak{M}$.

DEFINITION 1. *The mapping $\mathcal{K}$ from $\mathfrak{M}$ into the set of all subsets of $\mathcal{L}(\mathfrak{M})$ is a bounded K-mapping iff*:

a) *The set $\mathcal{K}(\mathbf{M})$ is non-void and convex for every $\mathbf{M} \in \mathfrak{M}$.*

b) *If $\{\mathbf{M}_n\}$, $\{\mathbf{L}_n\}$ are sequences of matrices $\mathbf{M}_n \in \mathfrak{M}$, $\mathbf{L}_n \in \mathcal{K}(\mathbf{M}_n)$; $(n = 1, 2, \ldots)$ such that*

$$\lim_{n\to\infty} \mathbf{M}_n = \mathbf{M} \in \mathfrak{M},$$

$$\lim_{n\to\infty} \mathbf{L}_n = \mathbf{L} \in \mathcal{L}(\mathfrak{M}),$$

then $\mathbf{L} \in \mathcal{K}(\mathbf{M})$.

c) $\sup \{\|\mathbf{L}\| : \mathbf{L} \in \mathcal{K}(\mathbf{M}), \mathbf{M} \in \mathfrak{M}\} < \infty$.

EXERCISE V.2. Prove that each continuous mapping from $\mathfrak{M}$ into a bounded subset of $\mathcal{L}(\mathfrak{M})$ is a bounded K-mapping.

HINT. Use that a one-point set is convex and bounded.

PROPOSITION V.10. *Suppose that $\nabla\phi(\mathbf{M})$ is defined on $\mathfrak{M}_+$. Then the mapping $\mathcal{K}_0$ defined by*

$$\mathcal{K}_0(\mathbf{M}) \equiv \{\mathbf{M}(\xi) - \mathbf{M} : \mathbf{M}(\xi) \in \mathfrak{M}, \boldsymbol{f}'(\bar{x})\nabla\phi(\mathbf{M})\boldsymbol{f}(\bar{x}) = \min_{x\in X} \boldsymbol{f}'(x)\nabla\phi(\mathbf{M})\boldsymbol{f}(x); (\bar{x} \in X_\xi)\} \quad \text{if } \mathbf{M} \in \mathfrak{M}_+$$

$$\mathcal{K}_0(\mathbf{M}) \equiv \{\bar{\mathbf{M}} - \mathbf{M} : \bar{\mathbf{M}} \in \mathfrak{M}_+\} \quad \text{if } \mathbf{M} \in \mathfrak{M} - \mathfrak{M}_+ \tag{31}$$

is a bounded K-mapping.

PROOF. a) From Eq. (60) in Chapter IV it follows that the equalities

$$\boldsymbol{f}'(\bar{x})\nabla\phi(\mathbf{M})\boldsymbol{f}(\bar{x}) = \min_{x\in X} \boldsymbol{f}'(x)\nabla\phi(\mathbf{M})\boldsymbol{f}(x); \quad (\bar{x} \in X_\xi)$$

are equivalent to

$$\partial\phi(\mathbf{M}, \mathbf{M}(\xi)) = \min_{\bar{\mathbf{M}} \in \mathfrak{M}} \partial\phi(\mathbf{M}, \bar{\mathbf{M}}). \tag{32}$$

For any $\mathbf{M}\in\mathfrak{M}_+$ the function

$$\bar{\mathbf{M}}\in\mathfrak{M}\mapsto \mathrm{tr}\, \nabla\phi(\mathbf{M})(\bar{\mathbf{M}}-\mathbf{M})=\partial\phi(\mathbf{M},\bar{\mathbf{M}})$$

is continuous and linear on the compact set $\mathfrak{M}$. Hence it achieves its minimum on $\mathfrak{M}$, and the set

$$\mathscr{K}_0(\mathbf{M})=\{\bar{\mathbf{M}}-\mathbf{M}: \partial\phi(\mathbf{M},\bar{\mathbf{M}})=\min_{\mathbf{N}\in\mathfrak{M}}\partial\phi(\mathbf{M},\mathbf{N})\} \tag{33}$$

is convex and closed.

b) Suppose that $\mathbf{M}_n\in\mathfrak{M}_+$, $\bar{\mathbf{M}}_n-\mathbf{M}_n\in\mathscr{K}_0(\mathbf{M}_n)$, $\mathbf{M}_n\to\mathbf{M}_\infty\in\mathfrak{M}_+$, $\bar{\mathbf{M}}_n\to\bar{\mathbf{M}}_\infty$. According to Eq. (32), for every n

$$\partial\phi(\mathbf{M}_n,\bar{\mathbf{M}}_n)\leqslant\partial\phi(\mathbf{M}_n,\mathbf{M});\quad (\mathbf{M}\in\mathfrak{M}). \tag{34}$$

Having made use of Eqs. (60) and (61) in Chapter IV, we can write

$$\partial\phi(\mathbf{M},\bar{\mathbf{M}})=\mathrm{tr}\,\nabla\phi(\mathbf{M}|\mathscr{L}(\mathfrak{M}))(\bar{\mathbf{M}}-\mathbf{M});$$
$$(\mathbf{M},\bar{\mathbf{M}}\in\mathfrak{M}).$$

According to Proposition IV.25, the mapping

$$\mathbf{M}\in\mathscr{U}_\phi\mapsto\nabla\phi(\mathbf{M}|\mathscr{L}(\mathfrak{M}))$$

is continuous, hence the mapping

$$(\mathbf{M},\bar{\mathbf{M}})\in\mathfrak{M}_+\times\mathfrak{M}\mapsto\partial\phi(\mathbf{M},\bar{\mathbf{M}}) \tag{35}$$

is continuous as well. Taking the limit for $n\to\infty$, we obtain from Eq. (34)

$$\partial\phi(\mathbf{M}_\infty,\bar{\mathbf{M}}_\infty)\leqslant\partial\phi(\mathbf{M}_\infty,\mathbf{M});\quad (\mathbf{M}\in\mathfrak{M})$$

which means that

$$\bar{\mathbf{M}}_\infty-\mathbf{M}_\infty\in\mathscr{K}_0(\mathbf{M}_\infty). \tag{36}$$

Furthermore, Eq. (36) is satisfied obviously if $\mathbf{M}_\infty\in\mathfrak{M}-\mathfrak{M}_+$, or if the sequence $\{\mathbf{M}_n\}$ contains infinitely many members belonging to the set $\mathfrak{M}-\mathfrak{M}_+$.

c) For every $\mathbf{M}\in\mathfrak{M}$, $\bar{\mathbf{M}}-\mathbf{M}\in\mathscr{K}_0(\mathbf{M})$ there holds

$$\|\bar{\mathbf{M}}-\mathbf{M}\|\leqslant 2\sup\{\|\boldsymbol{f}(x)\|^2: x\in X\}<\infty.\ \square$$

PROPOSITION V,11 (*The continuity of* $\mathcal{K}$). *Let* $\mathcal{K}$ *be a bounded K-mapping defined on* $\mathfrak{M}$. *Take* $\mathbf{M}\in\mathfrak{M}$. *Then to every* $\varepsilon>0$ *there is* $\delta>0$ *such that for* $\bar{\mathbf{M}}\in\mathfrak{M}$ *there holds the implication*

$$(\|\mathbf{M}-\bar{\mathbf{M}}\|<\delta,\ \mathbf{L}\in\mathcal{K}(\mathbf{M}))\Rightarrow\varrho(\mathbf{L},\mathcal{K}(\bar{\mathbf{M}}))<\varepsilon. \tag{37}$$

Here

$$\varrho(\mathbf{L},\mathcal{K}(\bar{\mathbf{M}}))\equiv\inf\{\|\mathbf{L}-\bar{\mathbf{L}}\|:\bar{\mathbf{L}}\in\mathcal{K}(\bar{\mathbf{M}})\}.$$

PROOF. Suppose that the assertion is not true. In such a case there is an $\varepsilon>0$ such that for every n there are $\mathbf{M}_n\in\mathfrak{M}$, $\mathbf{L}_n\in\mathcal{L}(\mathfrak{M})$ such that $\|\mathbf{M}_n-\bar{\mathbf{M}}\|<1/n$, $\mathbf{L}_n\in\mathcal{K}(\mathbf{M}_n)$, and that

$$\varrho(\mathbf{L}_n,\mathcal{K}(\bar{\mathbf{M}}))\geqslant\varepsilon. \tag{38}$$

Denote

$$d\equiv\sup\{\|\mathbf{L}\|:\mathbf{L}\in\mathcal{K}(\mathbf{M}),\ \mathbf{M}\in\boldsymbol{m}\}.$$

From compactness of the set $\{\mathbf{L}\in\mathcal{L}(\mathfrak{M}):\|\mathbf{L}\|\leqslant d\}$ there follows that there is a subsequence $\{\mathbf{L}_{n_k}\}_k$ of $\{\mathbf{L}_n\}_n$ and a matrix $\bar{\mathbf{L}}$, $\|\bar{\mathbf{L}}\|\leqslant d$ so that

$$\lim_{k\to\infty}\mathbf{L}_{n_k}=\bar{\mathbf{L}}.$$

From $\lim_{n\to\infty}\mathbf{M}_n=\bar{\mathbf{M}}$ we obtain

$$\bar{\mathbf{L}}\in\mathcal{K}(\bar{\mathbf{M}}).$$

On the other hand, from Eq. (38) and from the continuity of the function $\mathbf{L}\in\mathcal{L}(\mathfrak{M})\mapsto\varrho(\mathbf{L},\mathcal{K}(\bar{\mathbf{M}}))$ it follows that

$$\varrho(\bar{\mathbf{L}},\mathcal{K}(\bar{\mathbf{M}}))\geqslant\varepsilon,$$

which is a contradiction. □

DEFINITION V.2. *Let* $\mathcal{K}$ *be a bounded K-mapping defined on* $\mathfrak{M}$. *The iterative procedure associated with* $\mathcal{K}$ *is a sequence* $\{\xi_n\}_{n=0}^{\infty}$ *of designs such that*

a) $\mathbf{M}(\xi_0)\in\mathfrak{M}_+$,

b) $\xi_{n+1}=(1-\beta_n)\xi_n+\beta_n\varkappa_n$; $(n=0,1,\ldots)$, *where*

$$\beta_n \in (0, 1), \quad \sum_{n=0}^{\infty} \beta_n = \infty, \quad \lim_{n\to\infty} \beta_n = 0,$$

and where the designs $\varkappa_n$ *are chosen so that*

$$\mathbf{M}(\varkappa_n) - \mathbf{M}(\xi_n) \in \mathscr{K}[\mathbf{M}(\xi_n)]; \quad (n = 0, 1, \ldots).$$

To simplify the notation denote

$$t_n \equiv \sum_{i=0}^{n} \beta_i; \quad (n = 0, 1, \ldots).$$

DEFINITION V.3. *The mapping*

$$\boldsymbol{\tau}: \langle 0, \infty) \mapsto \mathfrak{M}$$

such that

$$\boldsymbol{\tau}(t) = \mathbf{M}(\xi_n) + (t - t_n) \frac{\mathbf{M}(\xi_{n+1}) - \mathbf{M}(\xi_n)}{\beta_n}; \tag{39}$$
$$(t \in \langle t_n, t_{n+1}))$$

is the trajectory of the iterative procedure.

Obviously, the mapping $\boldsymbol{\tau}$ is differentiable almost everywhere (i.e. the set $\{t: \mathrm{d}\boldsymbol{\tau}(t)/\mathrm{d}t$ does not exist$\}$ has zero Lebesgue measure).

Denote

$$D \equiv 2 \max \{\|\boldsymbol{f}(x)\|^2 : x \in X\}$$

and let $\mathscr{D}$ be the set of all mappings $\boldsymbol{\varphi}: \langle 0, \infty) \mapsto \mathfrak{M}$ that are Lipschitz bounded with the bound D, i.e.

$$\frac{\|\boldsymbol{\varphi}(t) - \boldsymbol{\varphi}(\bar{t})\|}{|t - \bar{t}|} \leqslant D; \quad (t, \bar{t} \in \langle 0, \infty)), \tag{40}$$

and such that $\boldsymbol{\varphi}(0) = \mathbf{M}(\xi_0)$.

The trajectory of an iterative procedure belongs to $\mathscr{D}$.

We present the following proposition without proof (cf. [109], Chapter V, § 2).

PROPOSITION V.12. *Each mapping* $\boldsymbol{\varphi} \in \mathscr{D}$ *is differentiable almost everywhere.*

PROPOSITION V.13. *Let $\{\boldsymbol{\tau}_n\}$ be a sequence of mappings belonging to $\mathscr{D}$. Then there is a limit point $\boldsymbol{\tau}^*$ of this sequence, i.e. there is a subsequence $\{\boldsymbol{\tau}_{n_k}\}$ such that on any bounded interval $I \subset \langle 0, \infty)$*

$$\lim_{k\to\infty} \sup_{t\in I} \|\boldsymbol{\tau}_{n_k}(t) - \boldsymbol{\tau}^*(t)\| = 0.$$

PROOF. The mappings $\boldsymbol{\tau}_1, \boldsymbol{\tau}_2, \ldots$ are bounded uniformly on $\langle 0, r\rangle$ for any $r > 0$, i.e.

$$\sup_{n\geq 1} \sup_{t\in\langle 0, r\rangle} \|\boldsymbol{\tau}_n(t)\| \leq \sup\{\|\mathbf{M}\| : \mathbf{M} \in \mathfrak{M}\} + \mathrm{D}r,$$

and they are uniformly continuous on $\langle 0, r\rangle$ in the sense of

$$\lim_{\delta\searrow 0} \sup\{\|\boldsymbol{\tau}_n(t) - \boldsymbol{\tau}_n(s)\| : t, s \in \langle 0, r\rangle,$$
$$|t-s| < \delta, n \geq 1\} = 0.$$

Hence the Arzelào theorem from functional analysis can be used (cf. [99], § 7.4). According to it, the sequence $\{\boldsymbol{\tau}_n\}$ has a subsequence $\{\boldsymbol{\tau}_n^{(r)}\}$ converging on $\langle 0, r\rangle$ to some $\boldsymbol{\tau}^{(r)}$ in the sense of

$$\lim_{n\to\infty} \sup_{t\in\langle 0, r\rangle} \|\boldsymbol{\tau}_n^{(r)}(t) - \boldsymbol{\tau}^{(r)}(t)\| = 0.$$

Obviously,

$$\boldsymbol{\tau}^{(r)}(t) = \boldsymbol{\tau}^{(s)}(t); \quad (t \in \langle 0, r\rangle \cap \langle 0, s\rangle).$$

Define $\boldsymbol{\tau}^*(t) \equiv \boldsymbol{\tau}^{(r)}(t)$ for $t \in \langle 0, r\rangle$, and by

$$\boldsymbol{\tau}_{n_k} \equiv \boldsymbol{\tau}_k^{(k)}; \quad (k = 1, 2, \ldots)$$

define the desired subsequence. □

DEFINITION V.4. *Let $\boldsymbol{\tau}$ be the trajectory of an iterative procedure. A limit trajectory of the procedure is any limit point of the sequence of mappings*

$$t \in \langle 0, \infty) \mapsto \boldsymbol{\tau}(t + t_n); \quad (n = 0, 1, 2, \ldots).$$

According to Proposition V.13, every iterative procedure associated with a bounded K-mapping has at least one limit trajectory.

PROPOSITION V.14. *A limit trajectory* $\boldsymbol{\tau}^*$ *of an iterative procedure associated with* $\mathcal{K}$ *is differentiable almost everywhere on* $\langle 0, \infty)$, *and*

$$\frac{d\boldsymbol{\tau}^*(t)}{dt} \in \mathcal{K}[\boldsymbol{\tau}^*(t)]$$

almost everywhere on $\langle 0, \infty)$.

PROOF. The differentiability follows from Proposition V.12, since $\boldsymbol{\tau}^* \in \mathcal{D}$. According to Proposition V.13 there is a sequence of mappings

$$t \in \langle 0, \infty) \mapsto \boldsymbol{\tau}_k(t) \equiv \boldsymbol{\tau}(t + t_{n_k}); \quad (k = 1, 2, \ldots)$$

such that $\lim_k \boldsymbol{\tau}_k(t) = \boldsymbol{\tau}^*(t)$ uniformly on every bounded interval.

Choose a point $\bar{t} \in \langle 0, \infty)$ which has the derivative $d\boldsymbol{\tau}^*(t)/dt$. Denote $\bar{\mathbf{M}} \equiv \boldsymbol{\tau}^*(\bar{t})$. Take $\varepsilon > 0$. According to Proposition V.11 there is $\beta > 0$ so that

$$\mathcal{K}(\mathbf{M}) \subset \{\mathbf{L}: \varrho(\mathbf{L}, \mathcal{K}(\bar{\mathbf{M}})) \leqslant \varepsilon\} \equiv \mathcal{K}_\varepsilon(\bar{\mathbf{M}}),$$

as soon as $\|\mathbf{M} - \bar{\mathbf{M}}\| < 2\beta$. Due to the continuity of $\boldsymbol{\tau}^*$ there are numbers $s_1 \in \langle 0, \bar{t})$, $s_2 \in \langle \bar{t}, \infty)$ such that for every $t \in \langle s_1, s_2 \rangle \equiv I$ there holds the inequality

$$\|\boldsymbol{\tau}^*(t) - \bar{\mathbf{M}}\| < \beta.$$

On the other hand, there is such k_0 that for $k \geqslant k_0$ we have

$$\|\boldsymbol{\tau}_k(t) - \boldsymbol{\tau}^*(t)\| < \beta; \quad (t \in I).$$

Hence

$$\|\boldsymbol{\tau}_k(t) - \bar{\mathbf{M}}\| < 2\beta; \quad (t \in I),$$

which implies

$$\mathcal{K}(\boldsymbol{\tau}_k(t)) \subset \mathcal{K}_\varepsilon(\bar{\mathbf{M}}); \quad (t \in I,\ k \geqslant k_0).$$

Choose $t \in \langle s_1, \bar{t} \rangle$. Denote by $\boldsymbol{\tau}_k(t_j)$; $(j = 1, \ldots, J)$ the vertices of the trajectory $\boldsymbol{\tau}_k$ corresponding to the points $t_j \in \langle t, \bar{t} \rangle$. Denote further

$$\Delta t_j \equiv t_{j+1} - t_j; \quad (j = 1, \ldots, J),$$

$$\Delta t_0 \equiv t_1 - t,$$

$$\boldsymbol{r}_k(t_j) \equiv \boldsymbol{\tau}_k(t_{j+1}) - \boldsymbol{\tau}_k(t_j); \quad (j=1, \ldots, J),$$

$$\boldsymbol{r}_k(t_0) \equiv \boldsymbol{\tau}_k(t_1) - \boldsymbol{\tau}_k(t).$$

Obviously,

$$\boldsymbol{\tau}_k(\bar{t}) - \boldsymbol{\tau}_k(t) \leq \sum_{j=1}^{J} \boldsymbol{r}_k(t_j)\Delta t_j. \tag{41}$$

The set $\mathcal{K}_\varepsilon(\bar{\mathbf{M}})$ is convex, and

$$\boldsymbol{r}_k(t_j) \in \mathcal{K}_\varepsilon(\bar{\mathbf{M}}).$$

Thus

$$\sum_{j=0}^{J} \frac{\Delta t_j}{\bar{t} - t} \boldsymbol{r}_k(t_j) \in \mathcal{K}_\varepsilon(\bar{\mathbf{M}}),$$

which implies

$$\frac{\boldsymbol{\tau}_k(t) - \boldsymbol{\tau}_k(\bar{t})}{t - \bar{t}} \in \mathcal{K}_\varepsilon(\bar{\mathbf{M}}). \tag{42}$$

We proceed similarly when $t > \bar{t}$, $t \in I$. Letting k tend to infinity we obtain

$$\frac{\boldsymbol{\tau}^*(t) - \boldsymbol{\tau}^*(\bar{t})}{t - \bar{t}} \in \mathcal{K}_\varepsilon(\bar{\mathbf{M}}); \quad (t \in I),$$

hence

$$\varrho\left(\frac{\boldsymbol{\tau}^*(t) - \boldsymbol{\tau}^*(\bar{t})}{t - \bar{t}}, \mathcal{K}(\bar{\mathbf{M}})\right) \leq \varepsilon; \quad (t \in I).$$

Letting t converge to $\bar{t}$ we obtain

$$\varrho\left(\left.\frac{\mathrm{d}\boldsymbol{\tau}^*(t)}{\mathrm{d}t}\right|_{t=\bar{t}}, \mathcal{K}(\bar{\mathbf{M}})\right) \leq \varepsilon$$

for every $\varepsilon > 0$. It follows that

$$\left.\frac{\mathrm{d}\boldsymbol{\tau}^*(t)}{\mathrm{d}t}\right|_{t=\bar{t}} \in \mathcal{K}(\bar{\mathbf{M}}). \ \square$$

Denote

$$\mathfrak{M}^* \equiv \{\mathbf{M}^* \in \mathfrak{M},\ \phi(\mathbf{M}^*) = \min_{\mathbf{M} \in \mathfrak{M}} \phi(\mathbf{M})\}.$$

A bounded K-mapping $\mathscr{K}$ is said to generate trajectories decreasing for ϕ iff for every $\boldsymbol{\varphi}$: $\langle 0, \infty) \mapsto \mathfrak{M}_+$, $\boldsymbol{\varphi} \in \mathscr{D}$ such that

$$\frac{\mathrm{d}\boldsymbol{\varphi}(t)}{\mathrm{d}t} \in \mathscr{K}[\boldsymbol{\varphi}(t)]; \quad (\text{i.e. } t \in \langle 0, \infty))$$

the function

$$t \in \{\bar{t}\colon \bar{t} \in \langle 0, \infty),\ \boldsymbol{\varphi}(\bar{t}) \notin \mathfrak{M}^*\} \mapsto \phi[\boldsymbol{\tau}(t)]$$

is decreasing.

Proposition V.15. *Let $\mathscr{K}$ be a bounded K-mapping defined on $\mathfrak{M}$ and generating trajectories which are decreasing for ϕ. Let $\{\xi_n\}_{n=0}^{\infty}$ be the iterative procedure associated with $\mathscr{K}$. Then either $\{\mathbf{M}(\xi_n)\}_{n=0}^{\infty}$ has a limit point in $\mathfrak{M} - \mathfrak{M}_+$ or*

$$\lim_{n\to\infty} \phi[\mathbf{M}(\xi_n)] = \min\ \{\phi(\mathbf{M})\colon \mathbf{M} \in \mathfrak{M}\}.$$

Proof. Denote $\mathbf{M}_n \equiv \mathbf{M}(\xi_n)$. Suppose that the statement is not true. Then all limit points of $\{\mathbf{M}_n\}_{n=0}^{\infty}$ belong to $\mathfrak{M}_+$ and at least one does not belong to $\mathfrak{M}^*$. Denote this limit point by $\mathbf{M}$. Denote the trajectory of the iterative procedure by $\boldsymbol{\tau}$. For an adequate choice of $n_1, n_2, \ldots$ we obtain

$$\mathbf{M} = \lim_{k\to\infty} \mathbf{M}_{n_k}, \tag{43}$$

$$\boldsymbol{\tau}^*(t) = \lim_{k\to\infty} \boldsymbol{\tau}(t + t_{n_k}); \quad (t \in \langle 0, \infty)),$$

where $\boldsymbol{\tau}^*$ is one of the limit trajectories of the iterative procedure. We have $\mathbf{M} = \boldsymbol{\tau}^*(0)$, since $\mathbf{M}_{n_k} = \boldsymbol{\tau}(t_{n_k})$. Using that $\boldsymbol{\tau}^*(0) \notin \mathfrak{M}^*$, $\mathfrak{M}^*$ is a closed set, and $\boldsymbol{\tau}^*$ is a continuous mapping, we obtain that $\boldsymbol{\tau}^*(s) \notin \mathfrak{M}^*$ for some $s > 0$. Denote $\bar{\mathbf{M}} \equiv \boldsymbol{\tau}^*(s)$. Since $\mathscr{K}$ generates decreasing trajectories, the function $\phi[\boldsymbol{\tau}^*(\cdot)]$ is decreasing on the interval $\langle 0, s\rangle$, hence

$$\min\{\phi(\mathbf{N}) : \mathbf{N} \in \mathfrak{M}\} < \phi(\bar{\mathbf{M}}) < \phi(\mathbf{M}) - \varepsilon \tag{44}$$

for some $\varepsilon > 0$. From the definition of the point $\bar{\mathbf{M}}$ it follows that

$$\lim_{k\to\infty} \boldsymbol{\tau}(s + t_{n_k}) = \bar{\mathbf{M}}.$$

Let $t_{s_k} \equiv \min\{t_n : n = 0, 1, 2, \ldots, t_n \geqslant s + t_{n_k}\}$. Evidently,

$$\lim_{k\to\infty} |t_{s_k} - (s + t_{n_k})| \leqslant \lim_{k\to\infty} \beta_{s_k} = 0.$$

Since $\boldsymbol{\tau} \in \mathscr{D}$, we can write

$$\|\mathbf{M}_{s_k} - \boldsymbol{\tau}(s - t_{n_k})\| < D\,|t_{s_k} - (s - t_{n_k})|.$$

Hence

$$\lim_{k\to\infty} \mathbf{M}_{s_k} = \bar{\mathbf{M}}. \tag{45}$$

According to Proposition IV.19, the function ϕ is continuous, hence from Eqs. (43)—(45) it follows that there is k_0 such that

$$\phi(\mathbf{M}_{n_k}) < \phi(\mathbf{M}_{s_k}) - \varepsilon/2; \quad (k \geqslant k_0). \tag{46}$$

Define

$$t_k^* \equiv \max\{t : t_{n_k} \leqslant t \leqslant t_{s_k},\ \boldsymbol{\tau}(t) = \boldsymbol{\tau}(t_{n_k}) \equiv \mathbf{M}_{n_k}\}. \tag{47}$$

From the continuity of $t \in \langle 0, \infty) \mapsto \phi[\boldsymbol{\tau}(t)]$ it follows that the function

$$t \in (t_k^*, t_{s_k}\rangle \mapsto \phi[\boldsymbol{\tau}(t)] - \phi[\boldsymbol{\tau}(t_k^*)]$$

is either positive or negative. It cannot be negative, since Eq. (46) implies

$$\phi[\boldsymbol{\tau}(t_{s_k})] = \phi(\mathbf{M}_{s_k}) > \phi(\mathbf{M}_{n_k}) = \phi[\boldsymbol{\tau}(t_k^*)]; \quad (k \geqslant k_0).$$

Thus

$$\phi[\boldsymbol{\tau}(t)] > \phi[\boldsymbol{\tau}(t_k^*)]; \quad (k \geqslant k_0,\ t \in \langle t_k^*, t_{s_k}\rangle). \tag{48}$$

On the other hand, Eq. (46) implies

$$\phi[\boldsymbol{\tau}(t_{s_k})] - \phi[\boldsymbol{\tau}(t_k^*)] = \phi[\mathbf{M}_{s_k}] - \phi[\mathbf{M}_{n_k}] > \varepsilon/2; \quad (k \geqslant k_0).$$

Hence the uniform continuity of ϕ on the compact set $\mathfrak{M}$ implies

$$\|\boldsymbol{\tau}(t_k^*)-\boldsymbol{\tau}(t_{s_k})\|>\delta; \quad (k\geqslant k_0)$$

for some $\delta>0$. A consequence of inequalities just obtained is that

$$t_{s_k}-t_k^*\geqslant\frac{\|\boldsymbol{\tau}(t_{s_k})-\boldsymbol{\tau}(t_k^*)\|}{D}>\frac{\delta}{D}; \ (k\geqslant k_0). \tag{49}$$

Eqs. (48) and (49) imply

$$\phi[\boldsymbol{\tau}(t+t_k^*)]>\phi[\boldsymbol{\tau}(t_k^*)]; \ (k\geqslant k_0, t\in\langle 0, \delta/D\rangle). \tag{50}$$

Define

$$m_k=\min\{n: t_n\geqslant t_k^*\}; \quad (k\geqslant k_0).$$

From $0\leqslant t_{m_k}-t_k^*\leqslant\beta_{m_k}$ and from the definition of the iterative procedure we obtain

$$\lim_{k\to\infty}(t_{m_k}-t_k^*)=0,$$

hence

$$\lim_{k\to\infty}[\phi(\boldsymbol{\tau}(t-t_{m_k}))-\phi(\boldsymbol{\tau}(t+t_k^*))]=0; \ (t\geqslant 0). \tag{51}$$

Denote by $\bar{\boldsymbol{\tau}}$ a limit point of the sequence of mappings

$$t\in\langle 0,\infty)\mapsto\boldsymbol{\tau}(t+t_{m_k}); \quad (k\geqslant k_0).$$

The mapping $\bar{\boldsymbol{\tau}}$ is a limit trajectory of the iterative procedure. According to the definition of the points $\{t_{m_k}\}_{k\geqslant k_0}$, $\bar{\boldsymbol{\tau}}$ is also a limit point of the sequence of mappings

$$t\in\langle 0,\infty)\mapsto\boldsymbol{\tau}(t+t_k^*); \quad (k\geqslant k_0).$$

From Eqs. (50) and (51) we obtain

$$\begin{aligned}\phi[\bar{\boldsymbol{\tau}}(0)]&=\lim_{k\to\infty}\phi[\boldsymbol{\tau}(t_k^*)]\\ &\leqslant\liminf_{k\to\infty}\phi[\boldsymbol{\tau}(t+t_k^*)]\\ &=\phi[\bar{\boldsymbol{\tau}}(t)]; \quad (t\in\langle 0,\delta/D\rangle).\end{aligned} \tag{52}$$

From Eqs. (43), (47) and (51) we obtain, in turn,

$$\begin{aligned}\bar{\tau}(0) &= \lim_{k\to\infty} \tau(0+t_k^*)\\ &= \lim_{k\to\infty} \mathbf{M}_{n_k}\\ &= \mathbf{M}.\end{aligned}$$

Thus, according to the assumption at the beginning of the proof and according to Eq. (52), the points $\bar{\tau}(0)$ and $\bar{\tau}(\delta/D)$ do not belong to the set $\mathfrak{M}^*$. According to Proposition V.14, $d\bar{\tau}(t)/dt \in \mathcal{K}(\bar{\tau}(t))$ for almost every $t\in\langle 0, \infty)$ but, according to Eq. (52), the function $t \mapsto \phi[\bar{\tau}(t)]$ does not decrease. This is a contradiction to the assumed properties of $\mathcal{K}$. □

Consider now an iterative procedure associated with the mapping $\mathcal{K}_0$ defined in Proposition V.10.

a) The starting design ξ_0 is such that $\det \mathbf{M}(\xi_0) \neq 0$.
b) At the $(n+1)$-st step

$$\xi_{n+1} = (1-\beta_n)\xi_n + \beta_n \xi_{x_n},$$

where

$$\beta_n \in (0, 1), \quad \sum_{n=0}^{\infty} \beta_n = \infty, \quad \lim_{n\to\infty} \beta_n = 0,$$

and where the function

$$x \in X \mapsto \mathbf{f}'(x) \nabla\phi[\mathbf{M}(\xi_n)]\mathbf{f}(x) \tag{53}$$

achieves its minimum at the point x_n.

EXERCISE V.3. Modify the procedure so that

$$\xi_{n+1} = (1-\beta_n)\xi_n + \beta_n \varkappa_n,$$

where $\varkappa_n$ is an arbitrary design concentrated on the points of maxima of the function defined in Eq. (53). Prove that this modified procedure is associated with $\mathcal{K}_0$ as well.

PROPOSITION V.16. *If $\nabla\phi(\mathbf{M})$ is defined on $\mathfrak{M}_+$, then either the iterative procedure associated with $\mathcal{K}_0$ is convergent,*

$$\lim_{n\to\infty} \phi[\mathbf{M}(\xi_n)] = \min \{\phi(\mathbf{M}) : \mathbf{M} \in \mathfrak{M}\},$$

or $+\infty$ *is a limit point of the sequence* $\{\phi[\mathbf{M}(\xi_n)]\}_{n=0}^{\infty}$.

PROOF. It is sufficient to prove that $\mathcal{K}_0$ generates trajectories decreasing for ϕ. Let φ: $\langle 0, \infty) \in \mathfrak{M}_{++}$, $\varphi \in \mathcal{D}$, $\varphi(t) \in \mathfrak{M}_+ - \mathfrak{M}^*$, $\mathrm{d}\varphi(t)/\mathrm{d}t \in \mathcal{K}_0[\varphi(t)]$. According to the definition of $\mathcal{K}_0$ (comparing with Eq. (33)), we have

$$\mathcal{K}_0[\varphi(t)] = \{\mathbf{M}(\xi) - \varphi(t) : \partial\phi[\varphi(t), \mathbf{M}(\xi)] = \min_{\mathbf{M}\in\mathfrak{M}} \partial\phi[\varphi(t), \mathbf{M}]\}.$$

From Proposition IV.26 it follows that

$$\min_{\mathbf{M}\in\mathfrak{M}} \partial\phi[\varphi(t), \mathbf{M}] < 0.$$

Having made use of the expression for the directional derivative in Eq. (60), Chapter IV, we obtain

$$0 > \partial\phi \left[\varphi(t), \varphi(t) + \frac{\mathrm{d}\varphi(t)}{\mathrm{d}t}\right] = \operatorname{tr} \nabla\phi[\varphi(t)] \frac{\mathrm{d}\varphi(t)}{\mathrm{d}t} = \frac{\mathrm{d}\phi[\varphi(t)]}{\mathrm{d}t},$$

hence the function $\phi[\varphi(\cdot)]$ is decreasing at the point t. □

In order to avoid the escape of $\phi[\mathbf{M}(\xi_n)]$ to infinity, the increase of $\nabla\phi$ should be limited at each step. For this reason we shall suppose that:

(i) For every $c > 0$ the gradient $\nabla\phi$ is Lipschitz bounded on the set $\mathfrak{M}_c \equiv \{\mathbf{M} : \mathbf{M} \in \mathfrak{M},\ \phi(\mathbf{M}) \leq c\}$, i.e. there is $K_c > 0$ so that

$$\frac{\nabla\phi(\mathbf{M}) - \nabla\phi(\bar{\mathbf{M}})}{\|\mathbf{M} - \bar{\mathbf{M}}\|} \leq K_c; \quad (\mathbf{M}, \bar{\mathbf{M}} \in \mathfrak{M}_c).$$

(ii) There are numbers c, $\bar{c}$ such that $\phi[\mathbf{M}(\xi_0)] < c$, $c < \bar{c} < \infty$, and that

$$\sup_n \beta_n \leq (c - \phi[\mathbf{M}(\xi_0)])/(K_{\bar{c}} D^2),$$

$$\sup_n \beta_n \leq (\bar{c} - c)/(P_{\bar{c}} D),$$

where

$$P_c = \sup \{\|\nabla\phi(\mathbf{M})\| : \mathbf{M} \in \mathfrak{M}_c\}.$$

Notice that the assumption (ii) is not an essential restriction of the values of β_n or ϕ. Namely, it is sufficient to consider an arbitrary sequence $\{\gamma_n\}_{n=0}^{\infty}$ converging to zero, such that $\sum_{n=0}^{\infty} \gamma_n = \infty$, and to define $\beta_n \equiv Q\gamma_n/\gamma_0$, where

$$Q = \min \{(c - \phi[\mathbf{M}(\xi_0)])/K_{\bar{c}}D^2,\\ (\bar{c} - c)/D(DK_{\bar{c}} + \|\nabla\phi(\mathbf{M}_0)\|)\}.$$

In fact, from (i) it follows that for every $\mathbf{M} \in \mathfrak{M}_{\bar{c}}$

$$\|\nabla\phi(\mathbf{M})\| \leq \|\mathbf{M} - \mathbf{M}_0\| K_{\bar{c}} + \|\nabla\phi(\mathbf{M}_0)\|\\ \leq DK_{\bar{c}} + \|\nabla\phi(\mathbf{M}_0)\|,$$

hence

$$P_{\bar{c}} \leq DK_{\bar{c}} + \|\nabla\phi(\mathbf{M}_0)\|.$$

PROPOSITION V.17 [64]. *Suppose that* $\nabla\phi$ *and* $\{\beta_n\}_{n=0}^{\infty}$ *satisfy the assumptions (i) and (ii). Then*

$$\phi[\mathbf{M}(\xi_n)] \leq c; \quad (n = 0, 1, \ldots).$$

PROOF: We proceed by induction with respect to n. Obviously, $\phi[\mathbf{M}(\xi_0)] \leq c$. Denote $\mathbf{M}_n = \mathbf{M}(\xi_n)$ and suppose that $\phi(\mathbf{M}_n) \leq c$. According to the assumption (ii)

$$\|\mathbf{M}_{n+1} - \mathbf{M}_n\| = \beta_n \|\mathbf{f}(x_n)\mathbf{f}'(x_n) - \mathbf{M}_n\|\\ \leq \beta_n D \leq (\bar{c} - c)/P_{\bar{c}}. \tag{54}$$

We shall prove that

$$\phi(\mathbf{M}_{n+1}) \leq \bar{c}.$$

In fact, if $\phi(\mathbf{M}) \leq c$, $\phi(\bar{\mathbf{M}}) \geq \bar{c} + \varepsilon$ for some $\varepsilon > 0$, then, having denoted $h(\beta) \equiv \phi[(1-\beta)\mathbf{M} + \beta\bar{\mathbf{M}}]$, we obtain

$$0 < \bar{c} + \varepsilon - c \leq h(1) - h(0) = \int_0^1 [\mathrm{d}h(\beta)/\mathrm{d}\beta]\, \mathrm{d}\beta$$

$$= \int_0^1 \langle \nabla\phi[(1-\beta)\mathbf{M}+\beta\bar{\mathbf{M}}], \bar{\mathbf{M}}-\mathbf{M}\rangle \, d\beta$$
$$\leq P_{\bar{c}} \|\bar{\mathbf{M}}-\mathbf{M}\|.$$

Hence the distance of $\bar{\mathbf{M}}$ from the set $\mathfrak{M}_c$ is

$$\inf\{\|\mathbf{M}-\bar{\mathbf{M}}\| : \mathbf{M}\in\mathfrak{M}_c\}$$
$$= \inf\{\|\mathbf{M}-\bar{\mathbf{M}}\| : \phi(\mathbf{M})\leq c\}$$
$$\geq (\bar{c}+\varepsilon-c)/P_{\bar{c}} > (\bar{c}-c)/P_{\bar{c}}.$$

Thus Eq. (54) implies that $\mathbf{M}_{n+1}\in\mathfrak{M}_{\bar{c}}$.

We shall prove now that

$$\phi[\mathbf{M}_{n+1}]\leq c.$$

We can write

$$\phi(\mathbf{M}_{n+1})-\phi(\mathbf{M}_n)$$
$$= \beta_n \int_0^1 \langle \nabla\phi[\mathbf{M}_n + t\beta_n(\mathbf{f}(x_n)\mathbf{f}'(x_n)-\mathbf{M}_n)]$$
$$-\nabla\phi(\mathbf{M}_n), [\mathbf{f}(x_n)\mathbf{f}'(x_n)-\mathbf{M}_n]\rangle \, dt$$
$$+\beta_n\langle\nabla\phi(\mathbf{M}_n), [\mathbf{f}(x_n)\mathbf{f}'(x_n)-\mathbf{M}_n]\rangle$$
$$\leq \beta_n K_{\bar{c}} \int_0^1 t\beta_n \|\mathbf{f}(x_n)\mathbf{f}'(x_n)-\mathbf{M}_n\|^2 \, dt$$
$$+\beta_n \min_{\bar{\mathbf{M}}\in\mathfrak{M}} \partial\phi(\mathbf{M}_n, \bar{\mathbf{M}}) \leq \beta_n^2 K_{\bar{c}} D^2 + \beta_n \partial\phi(\mathbf{M}_n, \mathbf{M}^*),$$

where $\phi(\mathbf{M}^*) = \min_{\mathbf{M}\in\mathfrak{M}} \phi(\mathbf{M})$. The last inequality follows from Proposition IV.26.

From the convexity of ϕ and from the definition of the directional derivative it follows that

$$\partial\phi(\mathbf{M}_n, \mathbf{M}^*) \leq \phi(\mathbf{M}^*)-\phi(\mathbf{M}_n)$$
$$\leq \phi(\mathbf{M}_0)-\phi(\mathbf{M}_n).$$

Hence, making use of (ii)

$$\phi(\mathbf{M}_{n+1})-\phi(\mathbf{M}_n) \leq \beta_n^2 K_{\bar{c}} D^2$$
$$+\beta_n[\phi(\mathbf{M}_0)-\phi(\mathbf{M}_n)] \leq \beta_n[c-\phi(\mathbf{M}_n)]$$

thus

$$\phi(\mathbf{M}_{n+1}) \leq (1-\beta_n)\phi(\mathbf{M}_n) + \beta_n c \leq c. \square$$

COROLLARY. *If ϕ satisfies the assumptions (A) and (B) from Chapter IV.4, $\nabla\phi$ is defined on $\mathfrak{M}_+$ and it satisfies the assumption (i), and $\beta_n \in (0, 1)$, $\lim_{n\to\infty} \beta_n = 0$, $\sum_{n=0}^{\infty} \beta_n = \infty$ so that there holds (ii), then*

$$\lim_{n\to\infty} \phi[\mathbf{M}(\xi_n)] = \min \{\phi(\mathbf{M}) : \mathbf{M} \in \mathfrak{M}\}.$$

The stopping rule for the iterative procedure described is given in Proposition IV.28.

V.6. OTHER METHODS FOR COMPUTING ϕ-OPTIMUM DESIGNS

V.6.1. *The Steepest Descent Method with a Linear or Quadratic Iteration*

Consider a criterion function, which

a) has the properties (A) and (B) from Chapter IV.4,

b) for arbitrary $\mathbf{A}, \mathbf{B}, \mathbf{C} \in \mathcal{R}^{m\times m}$, $\mathbf{M} \in \mathfrak{M}_+$ has continuous derivatives

$$\begin{aligned} &\partial\phi(\mathbf{M}+\beta\mathbf{A})/\partial\beta|_{\beta=0}, \\ &\partial^2\phi(\mathbf{M}+\beta\mathbf{A}+\gamma\mathbf{B})/\partial\beta\partial\gamma|_{\beta=\gamma=0}, \\ &\partial^3\phi(\mathbf{M}+\beta\mathbf{A}+\gamma\mathbf{B}+\delta\mathbf{C})/\partial\beta\partial\gamma\partial\delta|_{\beta=\gamma=\delta=0} \end{aligned} \tag{55}$$

which are linear in the variables $\mathbf{A}, \mathbf{B}, \mathbf{C}$,

c) has the second-order derivative

$$\partial^2\phi(\mathbf{M}+\beta\mathbf{A})/\partial\beta^2|_{\beta=0} > 0 \tag{56}$$

for every $\mathbf{M} \in \mathfrak{M}_+$, $\mathbf{A} \in \mathfrak{S}^{m\times m}$, $\mathbf{A} \neq 0$. The linearity of the function $\mathbf{A} \mapsto \partial\phi(\mathbf{M}+\beta\mathbf{A})/\partial\beta|_{\beta=0}$ implies the existence of $\nabla\phi(\mathbf{M})$, and the assumption c) implies the convexity of ϕ on $\mathfrak{M}_+$.

The algorithm with a linear iteration is [43]:

a) $\det \mathbf{M}(\xi_0) \neq 0$.

b) $\xi_{n+1} = (1-\beta_n)\xi_n + \beta_n \varkappa_n$,

where $\varkappa_n$ is a design supported by the points at which the function $x \in X \mapsto \mathbf{f}'(x)\nabla\phi[\mathbf{M}(\xi_n)]\mathbf{f}(x)$ achieves its minimum, and β_n is the point where, in turn,

$$\beta \in \langle 0, 1 \rangle \mapsto \phi[(1-\beta)\mathbf{M}(\xi_n) + \beta\mathbf{f}(x_n)\mathbf{f}'(x_n)]$$

achieves its maximum.

c) The stopping rule is given in Proposition IV.28.

The algorithm with a quadratic iteration (cf. [43]) (the generalized Newton method):

a) $\det \mathbf{M}(\xi_0) \neq 0$.

b) At the $(n+1)$-st step

$$\xi_{n+1} = \xi_n + \beta_n \eta_n. \tag{57}$$

Here x_n is the point of the minimum of $\mathbf{f}'(x)\nabla\phi[\mathbf{M}(\xi_n)]\mathbf{f}(x)$, and η_n is a measure supported by the set $X_{\xi_n} \cup \{x_n\}$ that is defined as follows:

(i) Let $z_1, \ldots, z_k$ be the elements of the set $X_{\xi_n} \cup \{x_n\}$. Denote the vector of the first-order derivatives by $\boldsymbol{h}$, and the matrix of the second-order derivatives by $\mathbf{H}$:

$$h_i \equiv \partial\phi[\mathbf{M}(\xi_n) + \beta\mathbf{f}(z_i)\mathbf{f}'(z_i)]/\partial\beta|_{\beta=0};$$
$$(i = 1, \ldots, k),$$

$$\{\mathbf{H}\}_{ij}$$
$$\equiv \partial^2\phi[\mathbf{M}(\xi_n) + \beta\mathbf{f}(z_i)\mathbf{f}'(z_i) + \gamma\mathbf{f}(z_j)\mathbf{f}'(z_j)]/\partial\beta\partial\gamma|_{\beta=\gamma=0};$$
$$(i, j = 1, \ldots, k).$$

Denote

$$\boldsymbol{\xi}_n \equiv (\xi_n(z_1, \ldots, \xi_n(z_k)),$$
$$\boldsymbol{\eta} \equiv (\eta(z_1), \ldots, \eta(z_k))'.$$

According to the Taylor formula, the function $\phi[\mathbf{M}(\xi_n + \eta)]$ can be approximated by the expression

$$\phi[\mathbf{M}(\xi_n)] + \boldsymbol{\eta}'\boldsymbol{h} + \frac{1}{2}\boldsymbol{\eta}'\mathbf{H}\boldsymbol{\eta}. \tag{58}$$

We search for $\boldsymbol{\eta}_n$ to minimize the quadratic form (58) subject to the constraint

$$\sum_{i=1}^{k} \eta(z_i) = 0.$$

After some manipulations we obtain

$$\boldsymbol{\eta}_n = \mathbf{H}^{-1}(-\boldsymbol{h} + \lambda \mathbf{1}),$$

where $\mathbf{1} = (1, \ldots, 1)'$, and where

$$\lambda \equiv \frac{\mathbf{1}'\mathbf{H}^{-1}\boldsymbol{h}}{\mathbf{1}'\mathbf{H}^{-1}\mathbf{1}}.$$

In case of $\eta_n(z_{io}) < 0$ and $\xi_n(z_{io}) = 0$, we define $\eta_n(z_{io}) = 0$ and minimize the form (58) once more but on the set $X_{\xi_n} \cup \{x_n\} - \{z_{io}\}$.

(ii) Choose β_n so as to achieve the minimum of

$$\phi[\mathbf{M}(\xi_n)] + \beta \sum_{i=1}^{k} \boldsymbol{f}(z_i)\boldsymbol{f}'(z_i)\eta_n(z_i),$$

and check whether

$$\xi_n(z_i) + \beta_n \eta_n(z_i) \geq 0; \quad (i = 1, \ldots, k).$$

If these inequalities do not hold, the return to (i) and minimize the form (58) again, but subject to a supplementary constraint

$$\left\| \sum_{j=1}^{k} \boldsymbol{f}(z_j)\boldsymbol{f}'(z_j)\eta(z_j) \right\| \xi_n(z_i) \geq -C\eta(z_i); \quad (i = 1, \ldots, k).$$

c) Stop the procedure according to Proposition IV.28.

The convergence of both the linear and the quadratic procedures to $\min\{\phi(\mathbf{M}) : \mathbf{M} \in \mathfrak{M}\}$ is proved in [43].

V.6.2. *A Method Based on the Inverse Equivalence Theorem*

In the "general equivalence theorem" (Proposition IV.26) equivalent conditions for the ϕ-optimality of a design $\xi^* \in \Xi$ are expressed through the directional derivatives $\partial\phi(\mathbf{M}(\xi^*), \mathbf{M})$ at the point $\mathbf{M}(\xi^*)$. In [61] an "inverse equivalence theorem" is presented, the ϕ-optimality of ξ^* being expressed by the directional derivatives $\partial\phi(\mathbf{M}, \mathbf{M}(\xi^*))$ at the points $\mathbf{M}\in\mathfrak{M}$, $\mathbf{M}\neq\mathbf{M}(\xi^*)$.

Assumptions on the function ϕ:

1. ϕ is defined, finite and convex on a set $\mathscr{U}$ which:
   a) is open in $\mathscr{L}(\mathfrak{M})$,
   b) contains the set $\mathfrak{M}_\phi \equiv \{\mathbf{M}\in\mathfrak{M}: \phi(\mathbf{M})<\infty\}$.
2. The gradient $\nabla\phi(\mathbf{M})$ is defined on $\mathscr{U}$.
3. If $\phi(\mathbf{M})=\infty$, then $\mathbf{M}\in\mathfrak{M}-\mathfrak{M}_+$.

PROPOSITION V.18 [61] (*the inverse equivalence theorem*). *Let* $\mathbf{M}^*\in\mathfrak{M}$. *The inequalities*

$$\mathrm{tr}\,[\mathbf{M}^*\nabla\phi(\mathbf{M})]\leq\mathrm{tr}\,[\mathbf{M}\nabla\phi(\mathbf{M})]; \quad (\mathbf{M}\in\mathfrak{M},\ \phi(\mathbf{M})\leq c) \tag{59}$$

hold for every $c>\inf\{\phi(\mathbf{M}): \mathbf{M}\in\mathfrak{M}\}$ *exactly if there is a sequence* $\{\mathbf{M}_n\}$ *of points of* $\mathfrak{M}$ *such that*

a) $\phi(\mathbf{M}_n)<\infty; \quad (n=1, 2, \ldots)$,

b) $\lim_{n\to\infty}\mathbf{M}_n=\mathbf{M}^*$,

c) $\lim_{n\to\infty}\phi(\mathbf{M}_n)=\inf\{\phi(\mathbf{M}): \mathbf{M}\in\mathfrak{M}\}$.

REMARKS. The inequalities (59) can be expressed as

$$\partial\phi(\mathbf{M}, \mathbf{M}^*)\leq 0; \quad (\mathbf{M}\in\mathfrak{M},\ \phi(\mathbf{M})\leq c)$$

(compare with Eq. (60) in Chapter IV). For some criteria functions, b) and c) imply

$$\lim_{n\to\infty}\phi(\mathbf{M}_n)=\phi(\mathbf{M}^*).$$

Proposition V.18 allows to minimize ϕ in solving a linear programme (generally infinite-dimensional). ϕ which has the properties (A) and (B) given in Chapter 4 leads to the search for the smallest nonnegative number ω so that

$$(\exists\bar{\mathbf{M}}\in\mathfrak{M})(\forall\mathbf{M}\in\mathfrak{M}_c)\ \mathrm{tr}\,\bar{\mathbf{M}}\nabla\phi(\mathbf{M})-\omega\leq\mathrm{tr}\,\mathbf{M}\nabla\phi(\mathbf{M}).$$

If $\omega=0$, then $\phi(\bar{\mathbf{M}})=\min\{\phi(\mathbf{M}): \mathbf{M}\in\mathfrak{M}\}$.

The drawback of this method (and of any method based on duality) is that that it is the optimum information matrix and not the design that we obtain.

V.6.3. *Combined Iterative Methods*

In case of a finite set $X = \{x_1, \ldots, x_k\}$, with k being not too large, the computation of ϕ-optimum designs reduces to the minimization of a convex function on a simplex:

$$\min \left\{ \phi[\mathbf{M}(\xi)] : \sum_{i=1}^{k} \xi(x_i) = 1, \xi(x_1) \geq 0, \ldots, \xi(x_k) \geq 0 \right\}.$$

In case of a large (or infinite) set X, it is advantageous to combine methods for small sets X with more general methods, e.g. with the method of the steepest descent. When combining in such a way, one procedure can be considered as an inner iteration (in smaller loops) and the other one as an outer iteration (in larger loops).

Suppose in this section that the function ϕ has the property (A) given in Chapter IV.4, hence it is a global criterion function, further suppose that $\nabla\phi(\mathbf{M})$ is defined for $\mathbf{M} \in \mathfrak{M}_+$.

If $X = \{x_1, \ldots, x_k\}$, we can use the following algorithm (cf. [63]):

1. Start with ξ_0 such that $\xi_0(x_i) > 0$; $(i = 1, \ldots, k)$.

2. Choose a symmetric positive definite $k \times k$ matrix $\mathbf{\Lambda}$. The iteration is

$$\xi_{n+1} = \xi_n + \beta_n \eta_n,$$

where

a) $\eta_n = (\mathbf{1}'\mathbf{\Lambda}\mathbf{1})(\mathbf{\Lambda}d) - (\mathbf{1}'\mathbf{\Lambda}d)(\mathbf{\Lambda}\mathbf{1})$,

$$\mathbf{d} \equiv (d_1, \ldots, d_k)', \quad d_i \equiv -\mathbf{f}'(x_i)\nabla\phi[\mathbf{M}(\xi_n)]\mathbf{f}(x_i),$$

b) β_n is the point of minimum of the function

$$\beta \in \langle 0, \bar{\beta} \rangle \mapsto \phi[\mathbf{M}(\xi_n + \beta\eta_n)],$$

where

$$\bar{\beta} = \min \left\{ \frac{\xi_n(x_i)}{|\eta_n(x_i)|} : \eta_n(x_i) < 0 \right\}.$$

If X is an infinite or a finite but large set, proceed as follows:

1. Choose $X_k \equiv \{x_1, \ldots, x_k\} \subset X$ and do the described iterative procedure on X_k (the inner iteration).

2. Stop the inner iterative procedure at the n-th step if the number

$$-\bar{\beta} \sum_{i=1}^{k} \eta_n(x_i) \mathbf{f}'(x_i) \nabla \phi[\mathbf{M}(\xi_n)] \mathbf{f}(x_i)$$

is small. This corresponds to the requirement that the directional derivative $\partial\phi[\mathbf{M}(\xi_n), \mathbf{M}(\xi_{n+1})]$ is near zero.

3. Stop the computation if the stopping rule from Proposition IV.28 is satisfied on the whole set X. Otherwise, define

$$X_{k+1} = X_k \cup \{x_n\},$$

where x_n is the point of the minimum of

$$x \in X \mapsto \mathbf{f}'(x) \nabla \phi[\mathbf{M}(\xi_n)] \mathbf{f}(x). \tag{60}$$

Define

$$\xi_{n+1} = (1 - \beta_n)\xi_n + \beta_n \xi_{x_n},$$

where β_n minimizes the expression $\phi[(1-\beta)\mathbf{M}(\xi_n) + \beta\mathbf{M}(\xi_{x_n})]$ (the outer iteration).

4. Do the inner iterative procedure on X_{k+1} with the starting design ξ_{n+1}.

In [63] it has been proven that $\{\phi[\mathbf{M}(\xi_n)]\}_{n=0}^{\infty}$ converges monotonically to $\min\{\phi(\mathbf{M}) : \mathbf{M} \in \mathfrak{M}\}$.

V.6.4. *The Computation of ϕ-optimum Designs under Constraints*

In some cases supplementary constraints are to be imposed on allowed designs. In such cases it is typical that the one-point designs ξ_x; $(x \in X)$ are excluded from consideration, which makes the use of Proposition IV.27 or Eq. (66) in Chapter IV impossible.

Such optimization under supplementary constraints arises for example when another alternative regression model is to be considered, and it is therefore necessary to maintain a certain level of significance of a test for testing the model.

In [68] a case is considered when $\phi(\mathbf{M}) = -\log\det\mathbf{M}$ (D-optimality) and the alternative model

$$E[y(x)] = \tilde{\mathbf{f}}'(x)\tilde{\boldsymbol{\alpha}} \equiv \mathbf{f}'(x)\boldsymbol{\alpha} + f_{m+1}(x)\alpha_{m+1}; \quad (x \in X)$$

differs from the original model $E[y(x)] = \mathbf{f}'(x)\boldsymbol{\alpha}$; $(x \in X)$ by one parameter α_{m+1}. To test the hypothesis $\alpha_{m+1} = 0$ against the alternative hypothesis $\alpha_{m+1} \neq 0$ it is necessary to keep the value of $\mathrm{var}_\xi \alpha_{n+1}$ below a certain level, say c, in the alternative model. We shall denote

$$\tilde{\mathbf{M}}(\xi) \equiv \sum_{x \in X} \tilde{\mathbf{f}}(x)\tilde{\mathbf{f}}'(x)\xi(x) = \begin{pmatrix} \mathbf{M}(\xi), & \tilde{\mathbf{M}}_{\mathrm{II}}(\xi) \\ \tilde{\mathbf{M}}'_{\mathrm{II}}(\xi), & \sum_{x \in X} f^2_{m+1}(x)\xi(x) \end{pmatrix},$$

$$\Psi(\tilde{\mathbf{M}}) \equiv \{\tilde{\mathbf{M}}^{-1}\}_{m+1,\, m+1} - c.$$

We have to solve the problem

$$\min\,\{\phi(\mathbf{M}) : \mathbf{M} \in \mathfrak{M}_+,\ \Psi[\tilde{\mathbf{M}}] \leq 0\}. \tag{61}$$

Define the penalty function for (61) as

$$L_r(\tilde{\mathbf{M}}, \lambda) \equiv \phi(\mathbf{M}) + \frac{1}{4r}\,[\varphi^2(\lambda + 2r\Psi(\tilde{\mathbf{M}})) - \lambda^2],$$

where $\varphi(t) = \max\{t, 0\}$.

The iterative procedure is the following:

(i) In the inner iteration, for given $\lambda^{(k)}$ and $r > 0$ we determine $\tilde{\mathbf{M}}^{(k)}$ so as to minimize $L_r(\tilde{\mathbf{M}}, \lambda^{(k)})$ on $\mathfrak{M}_+$. As proven in [68], the function

$$\mathbf{M} \in \mathfrak{M} \mapsto L_r(\mathbf{M}, \lambda^{(k)})$$

is convex on $\mathfrak{M}_+$ and has the properties (A) and (B) given in Chapter IV.4. It is differentiable on $\mathfrak{M}_+$ and has the Lipschitz property required in Proposition V.17. We use the iterative procedure associated with $\mathscr{H}_0$ (see Proposition V.10) and the stopping rule given in Proposition IV.28.

(ii) In the external iteration we set

$$\lambda^{(k+1)} = \lambda^{(k)} + 2r\Psi[\tilde{\mathbf{M}}^{(k)}],$$

where $\tilde{\mathbf{M}}^{(k)}$ is the matrix obtained in the inner iteration.

The convergence of the whole procedure is proved in [68]. It is based on the exposition in Chapter V.5.

V.6.5. *Complementary Notes*

1. The described iterative methods can only be applied to differentiable criteria functions. In case of G-optimality we can use the equivalence of D-optimum designs (Proposition IV.6) which holds when $\sigma^2(x) = \text{const}$.

2. In case of E-optimality we use that

$$\lim_{p\to\infty} \phi_p(\mathbf{M}) = \phi_\infty(\mathbf{M}); \quad \{\mathbf{M} \in \mathfrak{M}\} \tag{62}$$

(see Chapter IV.2.7), where

$$\phi_p(\mathbf{M}) = \left[\frac{1}{m} \operatorname{tr} \mathbf{M}^{-p}\right]^{1/p},$$

and where

$$\phi_\infty(\mathbf{M}) = [\lambda_{\min}(\mathbf{M})]^{-1}$$

is the E-optimality criterion function. We have

$$\phi_p(\mathbf{M}) \leq \phi_{p+1}(\mathbf{M}) \leq \phi_\infty(\mathbf{M}); \quad (\mathbf{M} \in \mathfrak{M}, \; p = 1, 2, \ldots). \tag{63}$$

Define $\mathbf{M}_p$, $\mathbf{M}_\infty$ by

$$\phi_p(\mathbf{M}_p) = \min\{\phi_p(\mathbf{M}) : \mathbf{M} \in \mathfrak{M}\},$$
$$\phi_\infty(\mathbf{M}_\infty) = \min\{\phi_\infty(\mathbf{M}) : \mathbf{M} \in \mathfrak{M}\}.$$

Take $c > \phi_\infty(\mathbf{M}_\infty)$. From Eq. (63) it follows that

$$\emptyset \subsetneqq \{\mathbf{M} : \phi_\infty(\mathbf{M}) \leq c\} \subset \{\mathbf{M} : \phi_p(\mathbf{M}) \leq c\} \subset \{\mathbf{M} : \phi_1(\mathbf{M}) \leq c\} \equiv Q,$$

hence $\mathbf{M}_\infty \in Q$, $\mathbf{M}_p \in Q$; $(p = 1, 2, \ldots)$. Moreover, according to Eq. (63), the functions $\phi_p(\cdot)$, $\phi_\infty(\cdot)$ are bounded above on the compact set Q by the number c. Hence, according to the Dinni theorem [110], the convergence (62) is uniform on Q. Consequently,

$$\lim_{p\to\infty} \phi_p(\mathbf{M}_p) = \phi_\infty(\mathbf{M}_\infty).$$

It follows that for a large p, $\mathbf{M}_p$ is well approximating the matrix $\mathbf{M}_\infty$.

3. For experiments with grouped observations (Chapter II.5.3), iterative methods analogous to those in Chapters V.2—V.6 can be used. We recall that the information matrix of a design ξ is equal to

$$\mathbf{M}_\mathbf{G}(\xi) = \sum_{x\in X} \mathbf{G}(x)\mathbf{K}^{-1}(x)\mathbf{G}'(x)\xi(x),$$

and that the variance of the BLUE for $\boldsymbol{g}'\boldsymbol{\alpha}$ is $\boldsymbol{g}'\mathbf{M}_\mathbf{G}^-(\xi)\boldsymbol{g}$. Thus again, the information matrix is linear in ξ, and we can do one-point corrections of designs, we can use an analogue of Proposition IV.28, etc.

4. In case of an incomplete model (see Chapter IV.5), we can complete it as in Chapter IV.5, and according to Proposition IV.32, any iterative method can be used in the complemented model.

Chapter VI

Design of Experiments in Particular Cases

Iterative methods presented in Chapter V are universal to a certain extent, but they need the use of computers and they do not help in gaining deeper insight into the structure of the optimum design. For some particular but important cases there are other, direct, noniterative methods of computation. We shall present some of them.

VI.1. THE REGRESSION MODEL LINEAR IN THE FACTORS

The regression model is said to be *linear in the factors* or a *first-order regression model* if

$$E_\alpha[y(t)] = \alpha_1 + \alpha_2 t_1 + \ldots + \alpha_m t_{m-1}; \quad (t \in X \subset R^{m-1}), \tag{1}$$

if $\sigma^2(t) = \text{const}$; $(t \in X)$ and X is a "regular geometrical formation" in R^{m-1}, e.g. a sphere, a simplex, a parallelepiped, the whole space R^{m-1}, etc. In this model we denote trials (i.e. elements of X) by $(m-1)$ dimensional vectors $\boldsymbol{t} \equiv (t_1, \ldots, t_{m-1})'$, $\boldsymbol{t}^{(1)}$, $\boldsymbol{t}^{(2)}$, etc., t_i being the "i-th factor" in the experiment.

Note that any model

$$E_\alpha[y(x)] = \boldsymbol{f}'(x)\boldsymbol{\alpha}; \quad (x \in X)$$

can be transformed into a linear form

$$E_\alpha[y(\boldsymbol{t})] = \boldsymbol{t}'\boldsymbol{\alpha}; \quad (\boldsymbol{t} \in \boldsymbol{f}(X))$$

by the transformation $x \in X \mapsto \boldsymbol{f}(x) \in R^m$; however, comparing with (1), the "absolute term" α_1 is omitted and the set $\boldsymbol{f}(X)$ can have a "nonregular" form.

A model which is linear in the factors is often used when there is no theoretical knowledge of the observed object, and the model is the simplest approximation to reality.

Denote by Ξ_N the set of designs that

a) are associated with a fixed size design $\mathbf{t}^{(1)}, \ldots, \mathbf{t}^{(N)}$ (see Definition III.3),

b) are "centred", i.e.

$$\sum_{t \in X} t_i \xi(\mathbf{t}) = 0; \quad (i = 1, \ldots, m-1),$$

c) are "normed", i.e.

$$\sum_{t \in X} t_i^2 \xi(\mathbf{t}) = 1; \quad (i = 1, \ldots, m-1).$$

PROPOSITION VI.1 [28]. *Suppose that X is a compact subset of R^{m-1} and that there is a design $\xi^* \in \Xi_N$ having a diagonal information matrix. Then*

$$1 = \operatorname{var}_{\xi^*} \alpha_i \leqslant \operatorname{var}_{\xi} \alpha_i; \quad (\xi \in \Xi_N, i = 1, \ldots, m) \tag{2}$$

and the design ξ^* is D-optimal and A-optimal within the set Ξ_N. Conversely, if Eq. (2) holds, then the matrix $\mathbf{M}(\xi^*)$ is diagonal.

PROOF [36]. Let $\xi \in \Xi_N$, det $\mathbf{M}(\xi) \neq 0$. There is a unique upper triangular matrix $\mathbf{W}$ (with zero entries below the diagonal) so that

$$\mathbf{M}(\xi) = \mathbf{W}\mathbf{W}'.$$

Then $\mathbf{W}^{-1}$ is triangular as well, and

$$\{\mathbf{W}^{-1}\}_{ii} = [\{\mathbf{W}\}_{ii}]^{-1}; \quad (i = 1, \ldots, m).$$

Since $\mathbf{M}^{-1}(\xi) = \mathbf{W}'^{-1}\mathbf{W}^{-1}$, we have

$$\begin{aligned} \{\mathbf{M}^{-1}(\xi)\}_{ii} &= \sum_{j=1}^{m} [\{\mathbf{W}^{-1}\}_{ji}]^2 \geqslant [\{\mathbf{W}^{-1}\}_{ii}]^2 \\ &= [\{\mathbf{W}\}_{ii}]^{-2} \geqslant \frac{1}{\sum_{j=1}^{m} [\{\mathbf{W}\}_{ij}]^2} = [\{\mathbf{M}(\xi)\}_{ii}]^{-1} = 1; \end{aligned} \tag{3}$$

$(i = 1, \ldots, m)$,

where we used that

$$\{\mathbf{M}(\xi)\}_{11}=\sum_{t}\xi(t)=1,$$

$$\{\mathbf{M}(\xi)\}_{ii}=\sum_{t}t_i^2\xi(t)=1; \quad (i=2,\dots,m).$$

From Eq. (3) it follows that

$$\sum_{j=1}^{m}[\{\mathbf{W}\}_{ij}]^2=[\{\mathbf{W}\}_{ii}]^2; \quad (i=1,\dots,m),$$

which implies that both $\mathbf{W}$ and $\mathbf{M}(\xi^*)$ are diagonal.

Conversely, if $\mathbf{M}(\xi^*)$ is diagonal, then Eq. (3) implies that $\mathrm{var}_{\xi^*}\alpha_i=1$, hence again from Eq. (3) it follows that $\mathrm{var}_{\xi^*}\alpha_i\leqslant\mathrm{var}_{\xi}\alpha_i$; $(i=1,\dots,m,$ $\xi\in\Xi_N)$. Thus Eq. (2) holds and ξ^* is A-optimal. Furthermore, from Eq. (3) we obtain that

$$\begin{aligned}\det\mathbf{M}(\xi)=[\det\mathbf{W}]^2=\prod_{i=1}^{m}[\{\mathbf{W}\}_{ii}]^2\leqslant 1\\ =\det\mathbf{M}(\xi^*); \quad (\xi\in\Xi_N)\end{aligned}$$

which is the D-optimality of ξ^* within the set Ξ_N. □

Consider a simplex in R^{m-1} having its centre at point $0\in R^{m-1}$, vertices (say $t^{(1)},\dots,t^{(m)}$) located symmetrically on the sphere centred at zero with the radius equal to $\sqrt{m-1}$. Consider the design ξ^* distributed uniformly on $t^{(1)},\dots,t^{(m)}$ (i.e. $\xi(t^{(i)})=1/m$). Clearly $\xi^*\in\Xi_m$. From the symmetry of the simplex it follows that

$$\sum_{i=1}^{m}t_j^{(i)}t_k^{(i)}=0; \quad (j\neq k).$$

Hence $\mathbf{M}(\xi^*)=\mathbf{I}$.

Making use of simplexes, designs optimum in the sense of Proposition VI.1 can be constructed. Denote by $\mathscr{V}$ the set of all simplexes in R^{N-1}, centred at zero and having vertices located symmetrically on the $(N-1)$-dimensional sphere

$$\{\mathbf{z}\colon \mathbf{z}\in R^{N-1}, \|\mathbf{z}\|^2=N-1\}.$$

PROPOSITION VI.2 (cf. [28]). *Suppose that* $X \subset R^{m-1}$ *and*

$$\{\boldsymbol{t}: \boldsymbol{t} \in R^{m-1}, \|\boldsymbol{t}\|^2 \leqslant N-1\} \subset X.$$

Then for $N \geqslant m$ *the information matrix of a design* $\xi \in \Xi_N$ *is diagonal exactly if* ξ *is distributed uniformly over the points which are obtained by projecting the N vertices of any simplex* $V \in \mathscr{V}$ *onto* $R^{m-1} \subset R^{N-1}$.

PROOF. Suppose first that $N = m$. As proven earlier, the information matrix of a design ξ distributed uniformly over the vertices $\boldsymbol{t}^{(1)}, \ldots, \boldsymbol{t}^{(m)}$ of V is diagonal. Conversely, suppose that

$$\mathbf{M}(\xi) = \mathbf{I}, \quad \xi(\boldsymbol{t}^{(1)}) = \ldots = \xi(\boldsymbol{t}^{(m)}) = 1/m.$$

We can write

$$\mathbf{M}(\xi) = \frac{1}{m} \mathbf{F}'\mathbf{F},$$

where $\mathbf{F}$ is an $m \times m$ matrix having the vector $(1, \boldsymbol{t}^{(i)})'$ on the i-th row. From the equality $\mathbf{F}'\mathbf{F} = m\mathbf{I}$ it follows that $\mathbf{F}\mathbf{F}' = m\mathbf{I}$. After decomposing the product $\mathbf{F}\mathbf{F}'$ we obtain

$$\|\boldsymbol{t}^{(i)}\|^2 = m-1$$
$$\boldsymbol{t}^{(i)\prime}\boldsymbol{t}^{(j)} = -1; \quad (i \neq j).$$

It follows that the points $\boldsymbol{t}^{(1)}, \ldots, \boldsymbol{t}^{(m)}$ are located equidistantly on

$$\{\boldsymbol{z}: \boldsymbol{z} \in R^m, \|\boldsymbol{z}\|^2 = \sqrt{m-1}\},$$

i.e. they are the vertices of V.

Suppose now that $N > m$. Let $\bar{\boldsymbol{t}}^{(1)}, \ldots, \bar{\boldsymbol{t}}^{(N)}$ be arbitrary points from R^{N-1} and let $\mathbf{F}^*$ be the matrix

$$\mathbf{F}^* \equiv \begin{pmatrix} 1, \bar{t}_1^{(1)}, \ldots, \bar{t}_{m-1}^{(1)}, \ldots, \bar{t}_{N-1}^{(1)} \\ \cdots\cdots\cdots\cdots\cdots\cdots\cdots \\ 1, \bar{t}_1^{(N)}, \ldots, \bar{t}_{m-1}^{(N)}, \ldots, \bar{t}_{N-1}^{(N)} \end{pmatrix}.$$

Similarly as above, $\frac{1}{N}\mathbf{F}^{*\prime}\mathbf{F}^*$ is an identity matrix exactly if the points $\bar{\boldsymbol{t}}^{(1)}, \ldots, \bar{\boldsymbol{t}}^{(N)}$ are vertices of a simplex $V \in \mathscr{V}$. Let us project the points $\bar{\boldsymbol{t}}^{(1)}$,

..., $\bar{t}^{(N)}$ on $R^{m-1} \subset R^{N-1}$. Denote by $t^{(1)}, \ldots, t^{(N)} \in R^{m-1}$ the obtained projections. Clearly

$$\bar{t}^{(i)\prime} = (t^{(i)}, \bar{t}_m^{(i)}, \ldots, \bar{t}_{N-1}^{(i)}); \quad (i = 1, \ldots, N).$$

Denote by $\mathbf{F}^{(0)}$ the $N \times N$ matrix

$$\mathbf{F}^{(0)} \equiv \begin{pmatrix} 1, t^{(1)\prime}, \mathbf{0} \\ \cdots\cdots\cdots\cdots \\ 1, t^{(N)\prime}, \mathbf{0} \end{pmatrix}.$$

Then

$$\mathbf{F}^{(0)\prime}\mathbf{F}^{(0)} = \begin{pmatrix} N\mathbf{M}(\xi), & \mathbf{0} \\ \mathbf{0}, & \mathbf{0} \end{pmatrix}.$$

Hence $\mathbf{M}(\xi)$ is diagonal, because $\mathbf{F}^{*\prime}\mathbf{F}^*$ is diagonal.

Conversely, if $\mathbf{M}(\xi) = \mathbf{I}$ and $X_\xi = \{t^{(1)}, \ldots, t^{(N)}\}$, then the first m columns of $\mathbf{F}^{(0)}$ are orthonormal. Set nonzero vectors instead of the zero columns of $\mathbf{F}^{(0)}$ so that the obtained matrix, say $\mathbf{F}^*$, would have only orthonormal columns. In such a case, as proven earlier, the columns of $\mathbf{F}^*$ are the vertices of the simplex V. □

The model linear in the factors having the set of trials defined by

$$X \equiv \{-1, 0, 1\}^{m-1} \tag{4}$$

is said to be the weighing model. In fact, the trial $t = (t_1, \ldots, t_{m-1})$ corresponds to the weighing of m objects on a balance, where $t_i = -1$ (or $t_i = 1$, or $t_i = 0$) denotes that the i-th object is on the left scale of the balance (or on the right scale, or it is not weighed) in the trial t.

PROPOSITION VI.3. *In the weighing model there hold the inequalities*

$$\mathrm{var}_\xi \alpha_i \geq 1; \quad (\xi \in \Xi, \ i = 1, \ldots, m). \tag{5}$$

Furthermore, $\mathrm{var}_\xi \alpha_i = 1$; $(i = 1, \ldots, m)$ *exactly if* $\mathbf{M}(\xi) = \mathbf{I}$. *Then the design* ξ *is A-optimal, D-optimal and G-optimal.*

PROOF. Suppose that $\det \mathbf{M}(\xi) \neq 0$. Similarly as in Proposition VI.1, denote by $\mathbf{W}$ the triangular matrix such that

$$\mathbf{M}(\xi) = \mathbf{W}\mathbf{W}'.$$

We can write

$$\operatorname{var}_\xi \alpha_i = \sum_{j=1}^{m} [\{\mathbf{W}^{-1}\}_{ji}]^2 \geq [\{\mathbf{W}^{-1}\}_{ii}]^2 = [\{\mathbf{W}\}_{ii}]^{-2}$$

$$\geq \left(\sum_{j=1}^{m} [\{\mathbf{W}\}_{ij}]^2\right)^{-1} = [\{\mathbf{M}(\xi)\}_{ii}]^{-1} \geq 1; \tag{6}$$

$(i = 1, \ldots, m)$.

The last inequality follows from

$$\{\mathbf{M}(\xi)\}_{ii} = \sum_{t \in X} t_i^2 \xi(t) \leq 1, \tag{7}$$

since t_i^2 is equal either to 1 or to 0.

The inequality in Eq. (6) is strict unless

$$\{\mathbf{M}(\xi)\}_{ii} = 1; \quad (i = 1, \ldots, m)$$

and

$$\sum_{j=1}^{m} [\{\mathbf{W}\}_{ij}]^2 = [\{\mathbf{W}\}_{ii}]^2; \quad (i = 1, \ldots, m),$$

which imply $\mathbf{M}(\xi) = \mathbf{I}$ and $\mathbf{W} = \mathbf{I}$.

In case of $\mathbf{M}(\xi^*) = \mathbf{I}$ the A-optimality of ξ^* is evident. D-optimality, and based on Proposition IV.6 also G-optimality, follow from

$$\det \mathbf{M}(\xi) = [\det \mathbf{W}]^2 = \prod_{i=1}^{m} [\{\mathbf{W}\}_{ii}]^2 \leq 1 = \det \mathbf{M}(\xi^*). \ \square$$

EXERCISE VI.1. Verify that $\mathbf{M}(\xi) = \mathbf{I}$ implies that in each trial t with $\xi(t) > 0$ all objects are on the balance.

The reader is referred to the book [29] for details on weighing experiments.

A particular case of the model linear in the factors is the model of the block experiment (see Chapter II.2, model (g)). The main difference compared with the analysis given earlier is that in the block experiment we are not interested in the "block parameters" but only in the "treatments", that is, partial optimality is considered. As shown in Exercise II.8, when $\alpha_1, \ldots, \alpha_v$ are the treatments, the covariance of the

BLUE-s of any linear functions of the treatments $\mathbf{h}'\boldsymbol{\alpha}$ and $\mathbf{g}'\boldsymbol{\alpha}$ is equal to

$$\mathrm{cov}_\xi(\mathbf{h}'\boldsymbol{\alpha}, \mathbf{g}'\boldsymbol{\alpha}) = \mathbf{h}'\mathbf{C}^- (\xi)\mathbf{g}.$$

The matrix $\mathbf{C}(\xi)$ is the "information matrix corresponding to the treatments", obtained from some entries of $\mathbf{M}(\xi)$ as explained in Exercise II.8. Similarly as in Schur's ordering of designs (see Chapter III.2), a partial Schur's ordering relative to the treatments is considered in several papers. We can roughly state that a design ξ^* is best in the sense of Schur's ordering relative to the treatments, if $\mathbf{C}(\xi^*)$ is completely symmetric (i.e. $\mathbf{C}(\xi^*) = a\mathbf{I} + b\mathbf{J}$, where $a, b \in R$, and $\mathbf{J}$ is the matrix of ones) and $\mathbf{C}(\xi^*)$ has a minimum trace and a minimum sum of entries within the class of $\mathbf{C}(\xi)$ matrices. Special results on Schur's optimality are obtained when the block sizes or the number of treatment replicas are fixed (the numbers r_i, k_j in Exercise II.8). For details cf. [37, 38].

VI.2. THE POLYNOMIAL REGRESSION ON THE REAL LINE

In this chapter we consider the regression experiment (X, Θ, σ) with

$$X = \langle a, b \rangle,$$

$$\Theta = \left\{\vartheta : \vartheta(x) = \sum_{i=1}^{m} \alpha_i x^{i-1};\ x \in \langle a, b \rangle, \boldsymbol{\alpha} \in R^m\right\}. \tag{8}$$

In this model the state function is a polynomial of order $m-1$. Attention has been paid mainly to the case of D-optimality (which is referred to as the optimum interpolation of a polynomial) and to the case of using the criterion function

$$\mathbf{M}(\xi) \in \mathfrak{M} \mapsto \mathrm{var}_\xi g_x,$$

where $x \in R - \langle a, b \rangle$ and g_x is, as already mentioned, the functional

$$g_x(\vartheta) = \sum_{i=1}^{m} \alpha_i x^{i-1};\quad \left(\vartheta(\cdot) = \sum_{i=1}^{m} \alpha_i (\cdot)^{i-1} \in \Theta\right)$$

(which is referred to as the optimum extrapolation of a polynomial from $\langle a, b\rangle$ to the point $x \notin \langle a, b\rangle$).

In this chapter the function $x \in X \mapsto \sigma(x)$ is no more supposed to be constant.

VI.2.1. *The Optimum Interpolation*

PROPOSITION VI.4. *If the D-optimum design ξ^* is supported by m points $\bar{x}_1, \ldots, \bar{x}_m \in \langle a, b\rangle$, then*

$$\xi^*(\bar{x}_i) = 1/m; \quad (i = 1, \ldots, m), \tag{9}$$

$$\prod_{k=1}^{m} \sigma^{-1}(\bar{x}_k) \left| \prod_{i<j}^{m} (\bar{x}_i - \bar{x}_j) \right|$$

$$= \max \left\{ \prod_{k=1}^{m} \sigma^{-1}(x_k) \left| \prod_{i<j}^{m} (x_i - x_j) \right| : x_1 < \ldots < x_m \right\}. \tag{10}$$

PROOF. We can write

$$\det \mathbf{M}(\xi^*) = (\det \mathbf{F})^2 \prod_{i=1}^{m} \sigma^{-2}(\bar{x}_i) \prod_{i=1}^{m} \xi^*(\bar{x}_i),$$

where

$$\mathbf{F} = \begin{pmatrix} 1, \bar{x}_1, \ldots, \bar{x}_1^{m-1} \\ \cdots\cdots\cdots\cdots \\ 1, \bar{x}_m, \ldots, \bar{x}_m^{m-1} \end{pmatrix}.$$

Computing the derivatives we can verify that the maximum of $\sum_{i=1}^{m} \log \xi(x_i)$, subject to $\sum_{i=1}^{m} \xi(x_i) = 1$, is achieved exactly if $\xi(x_i) = 1/m$; $(i = 1, \ldots, m)$. The determinant $\det \mathbf{F}$ is the determinant of Vandermond, thus there holds Eq. (10). □

PROPOSITION VI.5 [26]. *Let $\sigma^2(\cdot)$ be a polynomial positive on $\langle a, b\rangle$. Each D-optimum design is supported by m points from $\langle a, b\rangle$ if there is at least one of the following conditions satisfied.*

a) *For every $a \leq x_1 < x_2 < \ldots < x_{2m} = b$ we have*

$$\det \begin{pmatrix} \sigma^2(x_1), 1, x_1, \ldots, x_1^{2(m-1)} \\ \cdots\cdots\cdots\cdots\cdots\cdots \\ \sigma^2(x_{2m}), 1, x_{2m}, \ldots, x_{2m}^{2(m-1)} \end{pmatrix} \neq 0.$$

b) *The polynomial*

$$x \in \langle a, b \rangle \mapsto \frac{\mathrm{d}^{2m-1}\sigma^2(x)}{\mathrm{d}x^{2m-1}}$$

has no zero points on $\langle a, b \rangle$.

c) *The degree of the polynomial* $\sigma^2(\cdot)$ *is less or equal to* $2(m-1)$.

PROOF. Obviously, a D-optimum design cannot be supported by less than m points. Let ξ^* be a D-optimum design. Denote $\boldsymbol{f}(x) = (1, x, \ldots, x^{m-1})'$. Denote by Ψ the function

$$\Psi: x \in \langle a, b \rangle \mapsto m - \sigma^{-2}(x)\boldsymbol{f}'(x)\mathbf{M}^{-1}(\xi^*)\boldsymbol{f}(x).$$

According to Proposition IV.6

$$\Psi(x) \geq 0; \quad (x \in X),$$
$$\Psi(x) = 0; \quad (x \in X_{\xi^*}).$$

Evidently, $\Psi(x) = 0$ exactly if the polynomial $m\sigma^2(x) - \boldsymbol{f}'(x)\mathbf{M}^{-1}(\xi^*)\boldsymbol{f}(x) = 0$. The last polynomial has nonzero coefficients, hence Ψ has a finite number of zero points.

Suppose that the set X_{ξ^*} has at least $m+1$ points. They are zero points of Ψ, hence the equation

$$\Psi(x) - \gamma = 0 \tag{11}$$

has at least $2m$ solutions for some $\gamma > 0$.

The condition a) implies that for every $0 \leq x_1 < \ldots < x_{2m} \leq b$ the linear equations

$$\sigma^2(x_j)a_0 + \sum_{i=1}^{2m-1} a_i x_j^{i-1} = 0; \quad (j = 1, \ldots, 2m) \tag{12}$$

have a unique solution $a_0 = \ldots = a_{2m-1} = 0$. The degree of the polynomial $\boldsymbol{f}'(x)\mathbf{M}^{-1}(\xi^*)\boldsymbol{f}(x)$ being less than or equal to $2(m-1)$, according to Eq. (12) the polynomial

$$\varphi(x) \equiv (m - \gamma)\sigma^2(x) - \boldsymbol{f}'(x)\mathbf{M}^{-1}(\xi^*)\boldsymbol{f}(x)$$

has at most $2m-1$ zero points on $\langle a, b \rangle$. This is in contradiction with the proven properties of Ψ.

Since

$$\frac{d^{2m-1}\varphi(x)}{dx^{2m-1}}=(m-\gamma)\frac{d^{2m-1}\sigma^2(x)}{dx^{2m-1}},$$

from the condition b) it follows that the polynomial φ and hence also the function $\Psi-\gamma=\varphi\sigma^{-2}$ have at most $(2m-1)$ zero points on $\langle a, b\rangle$, which again is in contradiction with the properties of Ψ.

The condition c) implies that the degree of φ is at most $2(m-1)$, hence the function $\Psi-\gamma$ has at most $2(m-1)$ zero points. This is a contradiction as well. □

PROPOSITION VI.6 ([26] and [25]). *Let X and $\sigma^2(x)$ be defined by one of the formulae.*

1. $\sigma^2(x)=1$, $X=\langle -1, 1\rangle$,
2. $\sigma^2(x)=(1-x)^{\alpha-1}(1+x)^{\beta-1}$, $X=\langle -1, 1\rangle$, $\alpha<1$, $\beta<1$,
3. $\sigma^2(x)=e^x$, $X=\langle 0, \infty)$,
4. $\sigma^2(x)=x^{\alpha-1}e^x$, $X=\langle 0, \infty)$, $\alpha<1$,
5. $\sigma^2(x)=e^{x^2}$, $X=(-\infty, \infty)$.

Then there is a unique D-optimum design and it is supported by m zero points of one of the following polynomials

1. *the polynomial*

$$(1-x^2)\frac{dP_{m-1}(x)}{dx},$$

where P_n is the n-th Legendre polynomial (cf. [97]),

2. *the m-th Jacobi polynomial* (cf. [97]),
3. $xL_{m-1}^{(1)}(x)$, *where $L_n^{(\alpha)}(x)$ is the n-th Laguerre polynomial with parameter α* (cf. [97]),
4. *the polynomial $L_m^{(\alpha)}(x)$*,
5. *the m-th Hermite polynomial* (cf. [97]).

The proof is based on properties of orthogonal polynomials and can be found in [26].

An extension of the points 1 and 2 to the case of partial D-optimality is considered in [32] and [33].

VI.2.2. *The Optimum Extrapolation*

Now we shall consider designs which are optimal for the extrapolation of the polynomial $\vartheta(x)=\sum_{i=1}^{m}\alpha_i x^{i-1}$ from the interval $\langle a, b\rangle$ to a point $\bar{x}\in R-\langle a, b\rangle$.

Let $a\leqslant x_1<x_2<\ldots<x_m\leqslant b$. As above, $\boldsymbol{f}(x)=(1, x, \ldots, x^{m-1})'$. Denote

$$l_i(x)\equiv\frac{\det[\boldsymbol{f}(x_1), \ldots, \boldsymbol{f}(x_{i-1}), \boldsymbol{f}(x), \boldsymbol{f}(x_{i+1}), \ldots, \boldsymbol{f}(x_m)]}{\det[\boldsymbol{f}(x_1), \ldots, \boldsymbol{f}(x_m)]}$$

$$=\frac{\prod_{j\neq i}(x_j-x)}{\prod_{j\neq i}(x_j-x_i)}. \tag{13}$$

For every $x\in R$ the linear equation

$$\sum_{i=1}^{m} c_i\boldsymbol{f}(x_i)=\boldsymbol{f}(x)$$

has a unique solution $c_i=l_i(x)$; $(i=1, \ldots, m)$. Hence

$$\vartheta(x)=\sum_{i=1}^{m} l_i(x)\vartheta(x_i); \quad (\vartheta\in\Theta, x\in R). \tag{14}$$

The polynomials $l_1, \ldots l_m$ defined in Eq. (13) are the Lagrange interpolation polynomials. They will be used to construct designs minimizing $\mathrm{var}_\xi g_{\bar{x}}$.

Denote

$$\phi[\mathbf{M}(\xi)]=\mathrm{var}_\xi g_{\bar{x}}; \quad (\xi\in\Xi).$$

If $\det\mathbf{M}(\xi)\neq 0$, then (see Proposition II.7)

$$\phi[\mathbf{M}(\xi)]=\boldsymbol{f}'(\bar{x})\mathbf{M}^{-1}(\xi)\boldsymbol{f}(\bar{x}).$$

We can verify that

$$\nabla_{\mathbf{M}}\phi(\mathbf{M})=-\mathbf{M}^{-1}\boldsymbol{f}(\bar{x})\boldsymbol{f}'(\bar{x})\mathbf{M}^{-1},$$

$$f'(x)\nabla_{\mathbf{M}}\phi(\mathbf{M})f(x) = -[f'(x)\mathbf{M}^{-1}f(\bar{x})]^2; \quad (x \in R). \tag{15}$$

Suppose now that $X_\xi = \{x_1, \ldots, x_m\} \subset \langle a, b\rangle$. Using Eq. (14), we can write

$$f'(x)\mathbf{M}^{-1}(\xi)f(x^*) = \sum_{i=1}^{m} l_i(x)l_i(x^*)\sigma^2(x_i)/\xi(x_i); \quad (x, x^* \in R). \tag{16}$$

From Proposition IV.27 and from Eq. (15) it follows that a design μ, such that $\det \mathbf{M}(\mu) \neq 0$, minimizes $\mathrm{var}_\xi g_{\bar{x}}$ exactly if

$$\sigma^{-2}(x)[f'(x)\mathbf{M}^{-1}(\mu)f(\bar{x})]^2 \leq f'(\bar{x})\mathbf{M}^{-1}(\mu)f(\bar{x}); \quad (x \in \langle a, b\rangle) \tag{17}$$

with equality in (17) when $x \in X_\mu$.

PROPOSITION VI.7. *There is a design* ξ^* *minimizing* $\mathrm{var}_\xi g_{\bar{x}}$ *which is supported by an m-point set* $\{x_1, \ldots, x_m\} \subset \langle a, b\rangle$, *and*

$$\frac{\xi^*(x_i)}{\xi^*(x_j)} = \frac{|l_i(\bar{x})|\sigma(x_i)}{|l_j(\bar{x})|\sigma(x_j)}; \quad (i, j = 1, \ldots, m). \tag{18}$$

PROOF. According to Proposition III.16 there is ξ^* minimizing $\mathrm{var}_\xi g_{\bar{x}}$, which is supported by at most m points. However, if $X_{\xi^*} = \{x_1, \ldots, x_k\}$ with $k < m$, then

$$f(\bar{x}) = \sum_{i=1}^{k} c_i f(x_i),$$

since the estimate of $g_{\bar{x}}$ is unbiassed (see Propositions II.3 and II.7). But the last equation is not consistent, since from the properties of the vectors $f(x) \equiv (1, x, \ldots, x^{m-1})$; $(x \in X)$ it follows that the rank of the matrix $(f(\bar{x}), f(x_1), \ldots, f(x_k))$ is greater than the rank of $(f(x_1), \ldots, f(x_k))$. Therefore, ξ^* is supported by m points $x_1, \ldots, x_m \in \langle a, b\rangle$ and $\det \mathbf{M}(\xi^*) \neq 0$.

Denote $\mathbf{F} = (f(x_1), \ldots, f(x_m))$. Making use of Eq. (14), from Eq. (17)

we obtain

$$\sigma^{-2}(x_j)\left[\sum_{i=1}^{m}\{\mathbf{F}^{-1}\}_{j\cdot}\,\mathbf{f}(x_i)\,l_i(\bar{x})\sigma^2(x_i)/\xi^*(x_i)\right]^2=\text{const} \tag{19}$$

hence,

$$\begin{aligned}\text{const}&=\sigma^{-2}(x_j)[l_j(\bar{x})\sigma^2(x_j)/\xi^*(x_j)]^2\\&=\sigma^2(x_j)[l_j(\bar{x})/\xi^*(x_j)]^2.\end{aligned}$$

Thus Eq. (18) holds. □

PROPOSITION VI.8. *There is an* $(m-1)$*-degree polynomial p and there are points* $a\leqslant x_1<\ldots<x_m\leqslant b$ *such that*

a) $|p(x)|\leqslant\sigma(x)$,

b) $p(x_i)=(-1)^i\sigma(x_i)$; $(i=1,\ldots,m)$.

If $X_{\xi^*}=\{x_1,\ldots,x_m\}$ *and there holds Eq.* (18) *then*

$$\text{var}_{\xi^*}g_{\bar{x}}=\min_{\xi}\text{var}_{\xi}g_{\bar{x}}.$$

PROOF. According to Proposition VI.7 there is a design μ minimizing $\text{var}_\xi g_{\bar{x}}$ and concentrated on m points $x_1,\ldots,x_m\in\langle a,b\rangle$. From Eqs. (13), (16) and (18) it follows that

$$\begin{aligned}\mathbf{f}'(x)\mathbf{M}^{-1}(\mu)\mathbf{f}(\bar{x})&=\sum_{i=1}^{m}l_i(x)l_i(\bar{x})\sigma^2(x_i)/\mu(x_i)\\&=c\sum_{i=1}^{m}(-1)^i\sigma(x_i)\frac{\prod_{j\neq i}(x_j-x)}{\prod_{j\neq i}(x_j-x_i)},\end{aligned}$$

for every $x\in R$. Here $|c|=\sum_{j=1}^{m}|l_j(\bar{x})|\sigma(x_j)$. Denote

$$p(x)=\frac{1}{c}\,\mathbf{f}'(x)\mathbf{M}^{-1}(\mu)\mathbf{f}(\bar{x}).$$

Evidently,

$$p(x_i)=(-1)^i\sigma(x_i); \quad (i=1, \ldots, m)$$

and according to Eqs. (17) and (18) also

$$|p(x)| \leqslant \sigma(x); \quad (x \in \langle a, b \rangle).$$

Conversely, suppose that an $(m-1)$-degree polynomial has the properties a) and b). Let $l_1, \ldots, l_m$ be the Lagrange interpolation polynomials (Eq. (13)). Then, making use of Eq. (14),

$$\begin{aligned} p(x) &= \sum_{i=1}^{m} p(x_i) l_i(x) \\ &= \sum_{i=1}^{m} (-1)^i \sigma(x_i) l_i(x) \\ &= \sum_{i=1}^{m} \sigma^2(x_i) l_i(x) l_i(\bar{x}) / l_i(\bar{x}) \sigma(x_i). \end{aligned}$$

In defining

$$\xi^*(x_i) = \frac{|l_i(\bar{x})| \sigma(x_i)}{\sum_{j=1}^{m} |l_j(\bar{x})| \sigma(x_j)}; \quad (i=1, \ldots, m),$$

after having used Eq. (16), we obtain that

$$|p(x)| = \boldsymbol{f}'(x) \mathbf{M}^{-1}(\xi^*) \boldsymbol{f}(\bar{x}) \Big/ \sum_j |l_j(\bar{x})| \sigma(x_j); \quad (x \in R).$$

Thus the properties of the polynomial p and Eq. (16) imply that

$$\begin{aligned} &\sigma^{-2}(x)[\boldsymbol{f}'(x)\mathbf{M}^{-1}(\xi^*)\boldsymbol{f}(\bar{x})]^2 \\ &\leqslant \left[\sum_{j=1}^{m} |l_j(\bar{x})| \sigma(x_j)\right]^2 = \boldsymbol{f}'(\bar{x}) \mathbf{M}^{-1}(\xi^*) \boldsymbol{f}(\bar{x}) \end{aligned}$$

with equality if $x = x_i$; $(i=1, \ldots, m)$. Hence the condition (17) is verified, and

$$\operatorname{var}_{\xi^*} g_{\bar{x}} = \min_{\xi} \operatorname{var}_{\xi} g_{\bar{x}}. \ \square$$

COROLLARY. *Let* $\langle a, b\rangle = \langle -1, 1\rangle$, $\sigma(x)=1$; $(x\in\langle -1, 1\rangle)$ *and take* $\bar{x}\notin\langle -1, 1\rangle$. *Let*

$$x_i \equiv \cos\frac{(m-i)\pi}{m-1}; \quad (i=1, ..., m). \tag{20}$$

The design

$$\xi^*(x_i) = |l_i(\bar{x})| \Big/ \sum_{j=1}^{m} |l_j(\bar{x})|; \quad (i=1, ..., m) \tag{21}$$

is minimizing $\mathrm{var}_\xi g_{\bar{x}}$.

PROOF. From the identity

$$\cos k\varphi + \mathrm{i}\sin k\varphi = \mathrm{e}^{\mathrm{i}k\varphi} = (\cos\varphi + \mathrm{i}\sin\varphi)^k$$

we obtain that $\cos k\varphi$ is a polynomial in the variable $\cos\varphi$, which is of degree k. Thus

$$T(x) \equiv \cos[(m-1)\arccos x]; \quad x\in\langle -1, 1\rangle$$

is a polynomial of degree $m-1$. (The Tchebychef polynomial of the second kind, cf. [97].) Taking $x_1, ..., x_m$ as in Eq. (20), the conditions a) and b) in Proposition VI.8 can be verified. Finally, Eq. (21) is a consequence of Eq. (18). □

VI.3. THE TRIGONOMETRIC REGRESSION

Here we intend to call the reader's attention to the very simple situation which arises when we search for an optimum design in an experiment $(X, \Theta_k, 1)$ with

$$X = \langle 0, 2\pi\rangle,$$

$$\Theta_k = \left\{\vartheta: \vartheta(x) = \alpha_0 + \sum_{j=1}^{k} \alpha_j \cos jx + \sum_{j=1}^{k} \beta_j \sin jx;\ \alpha_j, \beta_j \in R\right\}.$$

The experiment $(X, \Theta_k, 1)$ is said to be the trigonometric regression of degree k.

PROPOSITION VI.9 [6]. *If* $n \geq 2k+1$, $X_\xi = \{x_1, x_2, \ldots, x_n\}$,

$$x_{j+1} - x_j = 2\pi/n; \quad (j=1, \ldots, n-1),$$
$$\xi(x_j) = 1/n; \quad (j=1, \ldots, n),$$

then ξ *is a D-optimum design.*

If μ *is D-optimal in the experiment* $(X, \Theta_k, 1)$, *then it is D-optimal in each experiment* $(X, \Theta_r, 1)$; $(r<k)$.

PROOF. Let us express the information matrix $\mathbf{M}(\xi)$. Obviously,

$$\{\mathbf{M}(\xi)\}_{00} = 1.$$

For $j \leq k$,

$$\{\mathbf{M}(\xi)\}_{jj} = \frac{1}{n} \sum_{l=1}^{n} \sin^2 j\left(x_1 + \frac{l-1}{n} 2\pi\right)$$
$$= \frac{1}{2} - \frac{1}{2n} \sum_{l=1}^{n} \cos 2j\left(x_1 + \frac{l-1}{n} 2\pi\right),$$

where $x_1 \in \langle 0, 2\pi/n \rangle$.

From the known properties of trigonometric functions we obtain for $\beta, \gamma \in R$

$$\sum_{l=1}^{n} \cos(\beta + (l-1)\gamma) + \mathrm{i} \sum_{l=1}^{n} \sin(\beta + (l-1)\gamma)$$
$$= \sum_{l=1}^{n} \mathrm{e}^{\mathrm{i}\beta}[\mathrm{e}^{\mathrm{i}\gamma}]^{l-1} = \mathrm{e}^{\mathrm{i}\beta} \frac{\mathrm{e}^{\mathrm{i}\gamma n} - 1}{\mathrm{e}^{\mathrm{i}\gamma} - 1}.$$

Hence

$$\sum_{l=1}^{n} \cos 2j\left(x_1 + \frac{l-1}{n} 2\pi\right)$$
$$= \sum_{l=1}^{n} \sin 2j\left(x_1 + \frac{l-1}{n} 2\pi\right) = 0.$$

It follows that

$$\{\mathbf{M}(\xi)\}_{jj} = \frac{1}{2}; \quad (j = 1, \dots, k).$$

Similarly,

$$\{\mathbf{M}(\xi)\}_{jj} = 1/2; \quad (j = k+1, \dots, 2k),$$
$$\{\mathbf{M}(\xi)\}_{ij} = 0; \quad (i \neq j, \ i, j = 0, \dots, 2k).$$

Thus $\mathbf{M}(\xi)$ is diagonal. Obviously, for every $x \in X$

$$\begin{aligned} \boldsymbol{f}'(x)\mathbf{M}^{-1}(\xi)\boldsymbol{f}(x) &= 1 + 2\sum_{j=1}^{k} \sin^2 jx + 2\sum_{j=1}^{k} \cos^2 jx \\ &= 1 + 2k \\ &= \text{the number of parameters.} \end{aligned}$$

So, according to Proposition IV.6, ξ is a D-optimum design.

The second statement in the proposition follows immediately. □

EXERCISE VI.2. Prove that the design ξ in Proposition VI.9 is G-optimal and A-optimal.

HINT. Show that the function

$$x \in \langle 0, 2\pi \rangle \mapsto \boldsymbol{f}'(x)\nabla\phi(\mathbf{M}(\xi))\boldsymbol{f}(x)$$

is constant, and use Propositions IV.6 and IV.27.

EXERCISE VI.3. Let λ be the Lebesgue measure on $\langle 0, 2\pi \rangle$ and let $\mu \equiv \lambda/2\pi$. Denote

$$\{\mathbf{M}(\mu)\}_{ij} \equiv \int_0^{2\pi} f_i(x) f_j(x) \mu(\mathrm{d}x).$$

Verify that the "design" μ is D-, A-, and G-optimal in the trigonometric regression experiment $(X, \Theta_k, 1)$.

HINT. Compare $\mathbf{M}(\mu)$ with $\mathbf{M}(\xi)$ in Proposition VI.9.

Chapter VII

The Functional Model and Measurements of Physical Fields

VII.1. INTRODUCTION

The regression model is applicable in measuring objects with a finite number of unknown parameters. However, there are important objects having states dependent on an infinite number of unknown parameters, namely the physical fields, such as the gravitational field, the electrostatic field, etc.

In this chapter a model of an experiment is described that is a direct generalization of the regression model with uncorrelated observations and that allows considering measurements of some physical fields. This model will be called here the functional model. The main difference when compared to the regression model is in the dimension of the set of states Θ. This seemingly unimportant difference has several consequences: the notion of a design must be enlarged, the derivatives of a state function $\vartheta \in \Theta$ are not estimable under any design, and instead of the matrix algebra used in the regression model methods of functional analysis, especially Hilbert space methods are to be used. On the other hand, results are obtained which allow a new insight also into the regression model.

We start the chapter with a short survey of the basic facts from the functional analysis. The reader is referred to [99, 100, 101, 102] for details.

We suppose that the reader is familiar with the definition of a *Hilbert space*. We recall the well-known *Schwarz inequality*: in a Hilbert space $\mathscr{H}$ with the inner product $\langle h_1, h_2 \rangle$ and the norm $\|h\|^2 \equiv \langle h, h \rangle$ there holds the inequality

$$|\langle h_1, h_2 \rangle| \leq \|h_1\| \, \|h_2\|; \quad (h_1, h_2 \in \mathcal{H}). \tag{1}$$

In Chapter VII.2 *the Riesz theorem* is used: if g is a linear bounded (i.e. continuous) functional defined on a Hilbert space $\mathcal{H}$, then there is a unique $h_g \in \mathcal{H}$ such that

$$g(h) = \langle h_g, h \rangle; \quad (h \in \mathcal{H}). \tag{2}$$

The norm of g is equal to

$$\|g\| \equiv \sup \left\{ \frac{|g(h)|}{\|h\|} : 0 \neq h \in \mathcal{H} \right\} = \|h_g\|. \tag{3}$$

Given a closed subspace $\mathcal{H}_0$ of $\mathcal{H}$, there is a unique linear operator P mapping $\mathcal{H}$ onto $\mathcal{H}_0$ such that

$$\begin{aligned} &\text{a)} \ \langle Ph_1, h_2 \rangle = \langle h_1, Ph_2 \rangle, \\ &\text{b)} \ PP = P. \end{aligned} \tag{4}$$

P is said to be *the orthogonal projector* onto $\mathcal{H}_0$. It is defined equivalently by

$$\|h - Ph\| = \min \{ \|h - l\| : l \in \mathcal{H}_0 \}.$$

When $\mathcal{H}_1$, $\mathcal{H}_2$ are two Hilbert spaces, by $\mathcal{H}_1 \oplus \mathcal{H}_2$ we denote the Hilbert space spanned by the set $\{(h_1, h_2): h_1 \in \mathcal{H}_1, \ h_2 \in \mathcal{H}_2\}$, and provided with the inner product

$$\langle (h_1, h_2), (h_1', h_2') \rangle \equiv \langle h_1, h_1' \rangle + \langle h_2, h_2' \rangle.$$

If $\{\mathcal{H}_n\}_{n=1}^{\infty}$ is a sequence of Hilbert spaces, then we denote by

$$\bigoplus_{n=1}^{\infty} \mathcal{H}_n$$

the Hilbert space of all sequences $\{h_i\}_{i=1}^{\infty}$; $(h_i \in \mathcal{H}_i)$ giving finite sums:

$$\sum_{i=1}^{\infty} \|h_i\|^2 < \infty.$$

The inner product in $\bigoplus_{n=1}^{\infty} \mathcal{H}_n$ is defined by

$$\langle \{h_i^{(1)}\}_{i=1}^{\infty}, \{h_i^{(2)}\}_{i=1}^{\infty} \rangle = \sum_{i=1}^{\infty} \langle h_i^{(1)}, h_i^{(2)} \rangle. \tag{5}$$

Two kinds of Hilbert spaces are used here at most: the reproducing kernel Hilbert space (RKHS) and the L^2-space.

Let T be a set and let

$$K: T \times T \mapsto R$$

be a function that is

a) symmetric, i.e. $K(t_1, t_2) = K(t_2, t_1)$,

b) positive semidefinite, i.e.

$$\sum_{i,j=1}^{n} c_i K(t_i, t_j) c_j \geq 0; \quad (c_1, \ldots, c_n \in R, \ t_1, \ldots, t_n \in T). \tag{6}$$

By $\mathcal{H}(K)$ we denote the RKHS with the kernel K, that is defined by the properties:

a) elements of $\mathcal{H}(K)$ are real-valued functions defined on T,

b) $K(\cdot, t) \in \mathcal{H}(K)$ for every $t \in T$,

c) given $t \in T$ and $h \in \mathcal{H}(K)$ we have

$$h(t) = \langle K(\cdot, t), h \rangle \tag{7}$$

(the "reproduction").

Let Z be a set. A σ-algebra of subsets of Z is a family $\mathcal{F}$ defined by the properties

a) $\emptyset \in \mathcal{F}$, $Z \in \mathcal{F}$,

b) $F_1, F_2 \in \mathcal{F} \Rightarrow F_1 - F_2 \in \mathcal{F}$, $F_1 \cup F_2 \in \mathcal{F}$,

c) $F_i \in \mathcal{F}$; $(i = 1, 2, \ldots) \Rightarrow \bigcup_{i=1}^{\infty} F_i \in \mathcal{F}$.

If Z is a topological space, then *the Borel σ-algebra* is the smallest σ-algebra containing all open subsets of Z. Particularly, if Z is a closed subset of R^n, then the Borel σ-algebra is spanned by the set

$$\{I \cap Z: I \in \mathcal{I}_n\},$$

$\mathcal{I}_n$ being the set of all n-dimensional intervals in R^n.

Let $(Z_1, \mathcal{F}_1)$, $(Z_2, \mathcal{F}_2)$ be two sets with σ-algebras. The mapping

$$\varphi: Z_1 \mapsto Z_2$$

is measurable iff

$$\{z_1: \varphi(z_1) \in F_2\} \in \mathcal{F}_1; \quad (F_2 \in \mathcal{F}_2).$$

Let Z be a set, $\mathscr{F}$ a σ-algebra of subsets of Z and η a probability measure on $\mathscr{F}$. By $L^2(Z, \mathscr{F}, \eta)$ (briefly: $L^2(\eta)$) we denote the Hilbert space of all equivalence classes of functions $l: Z \mapsto R$ that are square integrable,

$$\int_Z l^2 \, d\eta < \infty$$

(the L^2-space). Here, two functions l_1, l_2 are considered equivalent iff

$$\eta\{z: l_1(z) \neq l_2(z)\} = 0.$$

The inner product in $L^2(\eta)$ is defined by

$$\langle l_1, l_2 \rangle_\eta \equiv \int_Z l_1 l_2 \, d\eta. \tag{8}$$

Let y be a random variable defined on a set Ω, measurable with respect to a σ-algebra $\mathscr{S}$ of subsets of Ω ($\mathscr{S}$ = the σ-algebra of random events). Let P be a probability distribution on $\mathscr{S}$. Then y has a finite variance exactly if $y \in L^2(\Omega, \mathscr{S}, P)$. This follows from the inequalities:

$$|E_P(y)|^2 = |\langle y, 1 \rangle_P|^2 \leq \|y\|_P^2,$$
$$D_P(y) = \|y\|_P^2 - |\langle y, 1 \rangle_P|^2 \leq 2\|y\|_P^2,$$
$$\|y\|_P^2 = D_P(y) + |E_P(y)|^2.$$

Let $\mathscr{S}_0$ be a σ-algebra, $\mathscr{S}_0 \subset \mathscr{S}$. Then $L^2(\Omega, \mathscr{S}_0, P)$ is a subspace of the Hilbert space $L^2(\Omega, \mathscr{S}, P)$. The projector onto $L^2(\Omega, \mathscr{S}_0, P)$ is called *the conditional mean* and is denoted by $E_P(\cdot \mid \mathscr{S}_0)$.

Let Z be a set and let $\mathscr{F}$ be a σ-algebra of subsets of Z. A *bounded signed measure* on $\mathscr{F}$ is a function $v: \mathscr{F} \mapsto R$ having the properties:

a) $v(\emptyset) = 0$,

b) $v(F_1 \cup F_2) = v(F_1) + v(F_2)$; $(F_1 \cap F_2 = \emptyset)$,

c) if $F_n \searrow \emptyset$, then $\lim_{n \to \infty} v(F_n) = 0$,

d) $\sup\{|v(F)|: F \in \mathscr{F}\} < \infty$.

Moreover, if $v(F) \geq 0$; $(F \in \mathscr{F})$, then v is a bounded measure.

The Jordan-Hahn theorem: If v is a bounded signed measure, then there are bounded measures ν^+ and ν^- such that

$$\nu(F)=\nu^+(F)-\nu^-(F); \quad (F\in\mathscr{F}), \tag{9}$$

and there is a set $Z_+\subset Z$ such that $\nu^-(Z_+)=0$, $\nu^+(Z-Z_+)=0$. The measures ν^+ and ν^- are defined uniquely, the set Z_+ is defined uniquely up to a set of zero measure $|\nu|\equiv\nu^++\nu^-$.

A signed measure ν is said to be *absolutely continuous* with respect to a measure μ ($\nu\ll\mu$) iff

$$\mu(F)=0\Rightarrow\nu(F)=0; \quad (F\in\mathscr{F}).$$

In such a case there is an $\mathscr{F}$-measurable function denoted by $d\nu/d\mu$, having the property

$$\nu(F)=\int_F \frac{d\nu}{d\mu}\,d\mu; \quad (F\in\mathscr{F}).$$

The function $d\nu/d\mu$ is defined uniquely up to values on a set of zero measure μ.

If ν is a signed measure and μ is a measure, then there are signed measures ν_0, ν_1 such that

$$\nu=\nu_0+\nu_1,$$

ν_0 being absolutely continuous with respect to the measure μ, and ν_1 being orthogonal to μ. Consequently, there is a decomposition $Z=Z_1\cup Z_0$ such that $|\nu_1|(Z_0)=0$, $\mu(Z_1)=0$.

A normed linear space that is complete (i.e. every Cauchy sequence is convergent) is called a *Banach space*. Let B be a Banach space. Given any continuous (i.e. bounded) linear functional g defined on a subspace $B_0\subset B$, there is a linear functional g^* defined on B such that

$$g(b)=g^*(b); \quad (b\in B_0)$$

and

$$\begin{aligned}&\sup\left\{\frac{|g(b)|}{\|b\|}: 0\neq b\in B_0\right\}\\ &=\sup\left\{\frac{|g^*(b)|}{\|b\|}: 0\neq b\in B\right\}\end{aligned} \tag{10}$$

(*the Hahn-Banach theorem*).

Let X be a compact metric space. The set $C(X)$ of all functions defined and continuous on X is a Banach space with the norm

$$\|h\|_\infty \equiv \sup\{|h(x)| : x \in X\}. \tag{11}$$

The Riesz representation theorem for $C(X)$: given a continuous linear functional g on $C(X)$ there is a unique bounded signed measure ν defined on a Borel σ-algebra on X so that

$$g(h) = \int_X h \, d\nu; \quad (h \in C(X)). \tag{12}$$

VII.2. POTENTIAL FIELDS AND THE FUNCTIONAL MODEL

Consider the gravitational field generated by a mass point with mass m located at the point

$$\boldsymbol{x}^{(0)} \equiv (x_1^{(0)}, x_2^{(0)}, x_3^{(0)}) \in R^3.$$

The interaction of the field with objects located in the space can be determined from the function (*the potential of the field*)

$$\vartheta_m(\boldsymbol{x}) = -\frac{\varkappa m}{\|\boldsymbol{x} - \boldsymbol{x}^{(0)}\|}; \quad (\boldsymbol{x} \in R^3 - \{\boldsymbol{x}^{(0)}\}), \tag{13}$$

where $\varkappa$ is Newton's gravitational constant.

Basic physical variables derivable from the potential (13) are:

a) *The intensity* of the field

$$\boldsymbol{i}(\boldsymbol{x}) = (i_1(\boldsymbol{x}), i_2(\boldsymbol{x}), i_3(\boldsymbol{x})),$$

where

$$i_j(\boldsymbol{x}) = -\frac{\partial \vartheta(\boldsymbol{x})}{\partial x_j}; \quad (j = 1, 2, 3). \tag{14}$$

b) *The energy* of an object. If an object with mass density ϱ is located in the region V, then the energy necessary to remove the object from the field is

$$-\int_V \vartheta(\boldsymbol{x}) \varrho(\boldsymbol{x}) \, d\boldsymbol{x}. \tag{15}$$

c) *The total energy* of the field is proportional to the integral

$$\int_{R^3} \vartheta^2(\mathbf{x})\, d\mathbf{x}. \tag{16}$$

Another potential field, with a potential analogical to the one in Eq. (13), is the electrostatic field.

The potential generated by several mass points is the superposition (the sum) of potentials from the single points. Consequently, if the mass is distributed in a region V, according to a density ϱ, then at a point $\mathbf{x} \in R^3 - V$ the potential is equal to

$$\vartheta_\varrho(\mathbf{x}) = -\varkappa \int_V \frac{\varrho(\bar{\mathbf{x}})}{\|\mathbf{x} - \bar{\mathbf{x}}\|}\, d\bar{\mathbf{x}}. \tag{17}$$

Hence, from the localization of the sources of the field we can establish its potential, and consequently also other physical quantities in the field. However, in many experiments an inverse problem is solved, namely the potential or the intensity are measured and the sources generating the field are to be specified. Since measurements are often influenced by random fluctuations, a statistical model is necessary.

It is natural to identify the potential function of a field with the state of the field. Thus we obtain the set of states Θ. For example, if the source of a gravitational field is a mass point that has an unknown mass, then

$$\Theta = \{\vartheta_m : m > 0\} \tag{18}$$

with ϑ_m that has been defined in Eq. (1). The set Θ in Eq. (18) has the dimension equal to 1, hence an adequate model of an experiment for observing the field is the regression model.

A different situation arises if

$$\Theta = \{\vartheta_\varrho : \varrho = \text{bounded measure on } V\} \tag{19}$$

with ϑ_ϱ that has been defined in Eq. (17). In such a case Θ is an infinite-dimensional linear space, and the regression model cannot be used.

In Eq. (19) a state $\vartheta_\varrho(\cdot)$ is specified by a density function ϱ.

However, the set Θ can be described in different ways. According to [82], the potential ϑ of the gravitational field with sources localized in a region V can be characterized outside the set V by the following properties:

a) ϑ is a continuous twice differentiable function, which is a solution of the Laplace differential equation

$$\frac{\partial^2\vartheta}{\partial x_1^2}+\frac{\partial^2\vartheta}{\partial x_2^2}+\frac{\partial^2\vartheta}{\partial x_3^2}=0.$$

b) There is the limit

$$\lim_{\|\boldsymbol{x}\|\to\infty} \vartheta(\boldsymbol{x})=0.$$

c) There is a finite limit

$$\lim_{\|\boldsymbol{x}\|\to\infty} \|\boldsymbol{x}\|\vartheta(\boldsymbol{x}).$$

d) There are finite limits

$$\lim_{\|\boldsymbol{x}\|\to\infty} x_i^2\frac{\partial\vartheta(\boldsymbol{x})}{\partial x_i}; \quad (i=1, 2, 3).$$

Hence Θ is the set of functions ϑ: $R^3-V\mapsto R$ having the properties a)—d).

Another possibility is to define

$$\Theta=\mathcal{H}(\mathrm{K}),$$

where K is a RKHS with an adequately chosen kernel K (see Chapter VII.1). For example, according to [82], the set of possible gravitational potentials in space outside the Earth can be expressed like $\mathcal{H}(K)$ with the kernel K given by

$$K(\boldsymbol{x}, \tilde{\boldsymbol{x}})=\sum_{n=1}^{\infty}(2n+1)\frac{1}{\|\boldsymbol{x}\|\,\|\tilde{\boldsymbol{x}}\|}P_n\left(\frac{\boldsymbol{x}'\tilde{\boldsymbol{x}}}{\|\boldsymbol{x}\|\,\|\tilde{\boldsymbol{x}}\|}\right),$$

where P_n is the n-th Legendre polynomial [97],

$$P_n(t)=\frac{1}{2^n n!}\frac{d^n}{dt^n}(t^2-1)^n; \quad (t\in\langle -1, 1\rangle).$$

Denote by X the closed and bounded (i.e. compact) region of R^3 where the potential or the intensity of the field can be measured directly. Linear functionals which can typically be observed in a field are

1. $g_x: \vartheta \in \Theta \mapsto \vartheta(\mathbf{x})$; $(\mathbf{x} \in X)$,

2. $g_B^v: \vartheta \in \Theta \mapsto \int_B \vartheta(\mathbf{x}) v(d\mathbf{x})$; $(B \in \mathscr{B})$,

where $\mathscr{B}$ is the Borel σ-algebra on X and v is a bounded signed measure on X. In particular, the energy of an object localized in V with density ϱ is $-g_V^\varrho$.

3. $i_j: \vartheta \in \Theta \mapsto -\dfrac{\partial \vartheta(\mathbf{x})}{\partial x_j}$; $(j = 1, 2, 3;\ \mathbf{x} \in X)$,

4. $\vartheta \in \Theta \mapsto \int_B \dfrac{\partial \vartheta(\mathbf{x})}{\partial x_i} v(d\mathbf{x})$; $(B \in \mathscr{B})$,

where v is a signed measure on $\mathscr{B}$.

Resuming the discussion on potential fields the following definition is motivated (cf. [84, 85]).

DEFINITION VII.1. *The functional model of the first order is given by* $(X, \Theta, \sigma, \Sigma)$ *where*

X *is the closure of a bounded open subset of* R^n;

Θ *is an infinite-dimensional linear space of functions which are defined and continuously differentiable in the interior of* X;

σ *is a positive continuous function defined on* X;

$\Sigma: X \mapsto \mathscr{R}^{n \times n}$ *is a continuous mapping which associates a positive definite matrix* $\Sigma(x)$ *with each* $x \in X$.

A design in the model $(X, \Theta, \sigma, \Sigma)$ is given by two probability measures (ξ, η) defined on a sub-σ-algebra $\mathscr{B}_0$ of the Borel σ-algebra $\mathscr{B}$.

When the design (ξ, η) is chosen, we can observe random variables

$$y^\xi(B); \quad (B \in \mathscr{B}_0)$$

and random vectors

$$\mathbf{y}^{\eta}(B) \equiv (y_1^{\eta}(B), \ldots, y_n^{\eta}(B)); \quad (B \in \mathscr{B}_0)$$

such that for every $\vartheta \in \Theta$, $B_1, B_2 \in \mathscr{B}_0$

1. $\mathrm{E}_{\vartheta}[y^{\xi}(B_1)] = \int_{B_1} \vartheta(x)\, \mathrm{d}\xi(x),$

$$\operatorname{cov}[y^{\xi}(B_1), y^{\xi}(B_2)] = \int_{B_1 \cap B_2} \sigma^2(x)\, \mathrm{d}\xi(x),$$

$$y^{\xi}(B_1 \cup B_2) = y^{\xi}(B_1) + y^{\xi}(B_2) \text{ if } B_1 \cap B_2 = \emptyset.$$

2. $\mathrm{E}_{\vartheta}[y_i^{\eta}(B_1)] = \int_{B_1} \frac{\partial \vartheta(x)}{\partial x_i}\, \mathrm{d}\eta(x),$

$$\operatorname{cov}[y_i^{\eta}(B_1), y_j^{\eta}(B_2)] = \int_{B_1 \cap B_2} \{\Sigma(x)\}_{ij}\, \mathrm{d}\eta(x);$$

$(i, j = 1, \ldots, n),$

$$\mathbf{y}^{\eta}(B_1 \cup B_2) = \mathbf{y}^{\eta}(B_1) + \mathbf{y}^{\eta}(B_2) \text{ if } B_1 \cap B_2 = \emptyset.$$

3. $\operatorname{cov}[y_i^{\eta}(B_1), y^{\xi}(B_2)] = 0; \quad (i = 1, \ldots, n).$

REMARKS

A) Considering potential fields, the defined functional model corresponds to the "erroneous" observations of the potential and of the intensity of the field averaged over sets $B \in \mathscr{B}_0$.

When Θ is infinite-dimensional we cannot estimate the intensity of the field (i.e. the derivatives $\partial\vartheta/\partial x_i$) from the observations of the potential ϑ, because of the random errors. This is an important difference compared with the finite-dimensional regression model.

B) We could define the functional model of the s-th degree with $s > 1$, where partial derivatives of ϑ up to the s-th order could be observed directly.

C) The regression model (X, Θ, σ) is a functional model of zero order, with $\dim \Theta < \infty$. To make it evident, let ξ be a design in

(X, Θ, σ) associated with a fixed size design $x_1, \ldots, x_N$ (Definition II.3). Let us define

$$y^\xi(B) \equiv \frac{1}{N} \sum_{x_i \in B} y(x_i) = \sum_{x \in B} \bar{y}(x)\xi(x); \quad (B \in \mathscr{B}).$$

Evidently,

$$E_\vartheta[y^\xi(B)] = \int_B \vartheta(x) \, d\xi(x)$$

$$\operatorname{cov}[y^\xi(B_1), y^\xi(B_2)] = N \int_{B_1 \cap B_2} \sigma^2(x) \, d\xi(x).$$

It is not necessary to consider regression models of order $s \geqslant 1$. Indeed, if $\vartheta(x) = \sum_{i=1}^{m} \alpha_i f_i(x)$ and $f_1(\cdot), \ldots, f_m(\cdot)$ are differentiable functions, then

$$\frac{\partial \vartheta(x)}{\partial x_j} = \sum_{i=1}^{m} \alpha_i \frac{\partial f_i(x)}{\partial x_j},$$

hence the partial derivatives of ϑ are estimable under any nonsingular design.

D) Let $X = \langle 0, 1 \rangle$, $\Theta = C(\langle 0, 1 \rangle)$, $\sigma: \langle 0, 1 \rangle \mapsto 1$. Consider the functional

$$g: \vartheta \in C(\langle 0, 1 \rangle) \mapsto \int_0^1 \vartheta(x) \, dx.$$

From Proposition VII.2 it follows that g is estimable under a design ξ exactly if the measure ξ is absolutely continuous with respect to the Lebesgue measure on $\langle 0, 1 \rangle$. Hence it is not sufficient to consider designs supported by finite sets (compare with Definition II.3).

E) The functional model of the first order is composed of two relatively independent parts: the model (X, Θ, σ) (the direct generalization of the regression model) and the model (X, Θ, Σ), which could be considered a direct generalization of the regression model with grouped observations (see Chapter II.4.3).

We shall consider here only the model of order zero, (X, Θ, σ) with $\dim \Theta = \infty$, and $\sigma(x) = 1$; $(x \in X)$ (see the note in Chapter III.1).

VII.3. EXPERIMENTS FOR THE ESTIMATION OF LINEAR FUNCTIONALS

Let $(X, \Theta, 1)$ be a functional model (of order zero). Denote the set of all probability measures on $\mathcal{B}_0$ (= the set of designs) by $\bar{\Xi}$. Let $\xi \in \bar{\Xi}$. Denote $L^2(\xi) \equiv L^2(X, \mathcal{B}_0, \xi)$. Let $\mathcal{H}_\xi \subset L^2(\xi)$ be the Hilbert space spanned by the set $\{y^\xi(B): B \in \mathcal{B}_0\}$. Thus $\mathcal{H}_\xi$ contains the linear combinations of the observed random variables and their limits in the mean square. Therefore, $\mathcal{H}_\xi$ can be considered the space of linear estimates.

Suppose that the random variables $y^\xi(B)$; $(B \in \mathcal{B}_0)$ are defined on a basic space $(\Omega, \mathcal{S})$. Let P_ϑ be the probability distribution on $\mathcal{S}$ corresponding to the hypothesis that ϑ is the true state. The inner product in $L^2(\Theta, \mathcal{S}, P)$ will be denoted by $\langle y_1, y_2 \rangle_\vartheta \left(\equiv \int_\Omega y_1 y_2 \, \mathrm{d}P_\vartheta \right)$.

PROPOSITION VII.1. *There is a mapping*

$$\Psi \colon L^2(\xi) \mapsto \mathcal{H}_\xi$$

such that

a) *Ψ is a linear bijection,*

b) *under the hypothesis ϑ we have*

$$\mathrm{E}_\vartheta[\Psi(h)] = \int_X \vartheta h \, \mathrm{d}\xi; \quad (h \in \mathrm{L}^2(\xi)), \tag{20}$$

$$\mathrm{cov}_\vartheta[\Psi(h_1), \Psi(h_2)] = \int_X h_1 h_2 \, \mathrm{d}\xi; \tag{21}$$

$$(h_1 h_2 \in \mathrm{L}^2(\xi)).$$

In particular, if $\vartheta = 0$, then Ψ is an isomorphism of the Hilbert spaces $L^2(\xi)$ and $\mathcal{H}_\xi$.

PROOF. Let $\mathcal{L}$ (or $\mathcal{L}_y$) be the set of all finite linear combinations of the form

$$\sum_{i=1}^n \alpha_i \chi_{B_i} \quad \left(\text{or } \sum_{i=1}^n \alpha_i y^\xi(B_i) \right).$$

Define Ψ on $\mathscr{L}$ according to

$$\Psi\left(\sum_{i=1}^{n} \alpha_i \chi_{B_i}\right) = \sum_{i=1}^{n} \alpha_i y^{\xi}(B_i).$$

Thus Ψ is linear and $\Psi(\mathscr{L}) = \mathscr{L}_y$. If $\vartheta = 0$, then obviously

$$\mathrm{E}_0[\Psi(h)] = 0; \quad (h \in \mathscr{L}),$$

$$\begin{aligned} \operatorname{cov}[\Psi(h_1), \Psi(h_2)] &\equiv \langle \Psi(h_1), \Psi(h_2)\rangle_0 \\ &= \int_X h_1 h_2 \,\mathrm{d}\xi \equiv \langle h_1, h_2\rangle_{\xi}. \end{aligned}$$

Hence Ψ preserves the norm. Consequently, Ψ is one-to-one.

If $h \in \mathscr{H}_{\xi}$ ($\equiv$the closure of $\mathscr{L}$ in $\mathrm{L}^2(\xi)$) and if

$$\lim_{n\to\infty} \|h_n - h\|_{\xi} = 0$$

for $h_1, h_2, \ldots \in \mathscr{L}$, then $\Psi(h)$ is defined by

$$\lim_{n\to\infty} \|\Psi(h_n) - \Psi(h)\|_0 = 0.$$

This extension of Ψ from $\mathscr{L}$ to $\mathscr{H}_{\xi}$ is one-to-one, hence Ψ, when defined on $\mathscr{H}_{\xi}$, has the required properties under the hypothesis $\vartheta = 0$.

If $\vartheta \neq 0$, then

$$\begin{aligned} \|\Psi(h)\|_{\vartheta}^2 &= \mathrm{D}[\Psi(h)] + \mathrm{E}_{\vartheta}^2[\Psi(h)] \\ &= \|\Psi(h)\|_0^2 + \left[\int_X h\vartheta \,\mathrm{d}\xi\right]^2; \quad (h \in \mathscr{L}). \end{aligned}$$

Thus

$$\begin{aligned} \|\Psi(h)\|_0^2 &\leq \|\Psi(h)\|_{\vartheta}^2 \\ &\leq \|\Psi(h)\|_0^2 + \left[\int_X h^2 \,\mathrm{d}\xi \int_X \vartheta^2 \,\mathrm{d}\xi\right] \\ &= \|\Psi(h)\|_0^2 + \|\Psi(h)\|_0^2 \|\vartheta\|_{\xi}^2. \end{aligned}$$

It follows that the norms $\| \, \|_0$ and $\| \, \|_{\vartheta}$ are topologically equivalent, hence the extension of Ψ from $\mathscr{L}$ onto $\mathscr{H}_{\xi}$ does not depend on $\vartheta \in \Theta$. Evidently,

$$\mathrm{E}_\vartheta[\Psi(h)] = \int_X h\vartheta \, \mathrm{d}\xi; \quad (h \in \mathscr{L})$$

and $\mathrm{cov}_\vartheta[\Psi(h_1), \Psi(h_2)]$ does not depend on ϑ, which completes the proof. □

NOTE. Instead of $\Psi(h)$ we shall write the integral

$$\int_X h(x) y^\xi(\mathrm{d}x)$$

because $\Psi(h)$ has some properties of an integral. (Notice that in general it is not an integral, because the set function $B \in \mathscr{B}_0 \mapsto y^\xi(B, \omega)$ is only finitely additive for almost all $\omega \in \Omega$, $[P_\vartheta]$; cf. [87], Chapter 4.) From Proposition VII.1 it follows that

$$\int_X \alpha h(x) y^\xi(\mathrm{d}x) = \alpha \int_X h(x) y^\xi(\mathrm{d}x),$$

$$\int_X (h_1(x) + h_2(x)) y^\xi(\mathrm{d}x) = \int_X h_1(x) y^\xi(\mathrm{d}x)$$

$$+ \int_X h_2(x) y^\xi(\mathrm{d}x),$$

$$\mathrm{E}_\vartheta\left[\int_X h(x) y^\xi(\mathrm{d}x)\right] = \int_X h(x)\, \mathrm{E}_\vartheta[y^\xi(\mathrm{d}x)],$$

$$\lim_{n\to\infty} \|h_n - h\|_\xi = 0 \Rightarrow$$

$$\lim_{n\to\infty} \left\| \int_X h_n(x) y^\xi(\mathrm{d}x) - \int_X h(x) y^\xi(\mathrm{d}x) \right\|_\vartheta = 0.$$

In what follows we shall write briefly $\int h \, \mathrm{d}\xi$, $\int h(x) y^\xi(\mathrm{d}x)$ instead of

$$\int_X h \, \mathrm{d}\xi, \quad \int_X h(x) y^\xi(\mathrm{d}x).$$

Let

$$g\colon \Theta \mapsto R$$

be a linear functional. As in Chapter II.2, the functional g is said to be estimable (without bias) under the design ξ iff there is an $y \in \mathscr{H}_\xi$ such that

$$\mathrm{E}_\vartheta(y) = g(\vartheta); \quad (\vartheta \in \Theta).$$

The random variable y is an unbiassed estimate of g. The unbiassed estimate $\hat{g} \in \mathscr{H}_\xi$ is the BLUE iff its variance, $\mathrm{var}_\xi g$, is minimal. We define $\mathrm{var}_\xi g = \infty$ whenever g is not estimable.

Denote by P_ξ the orthogonal projection of $L^2(\xi)$ on $\bar{\Theta}^\xi \equiv$ the closure of Θ in $L^2(\xi)$.

PROPOSITION VII.2 [84]. *The functional g is estimable under ξ exactly if one of the following conditions is satisfied:*

a) $$g(\vartheta) = \int l\vartheta \, \mathrm{d}\xi; \quad (\vartheta \in \Theta)$$

for some $l \in L^2(\xi)$.

b) *There is a sequence $l_1, l_2, \ldots \in \Theta$ such that*

$$\lim_{m,\, n \to \infty} \int (l_n - l_m)^2 \, \mathrm{d}\xi = 0$$

and

$$g(\vartheta) = \lim_{n \to \infty} \int l_n \vartheta \, \mathrm{d}\xi.$$

c) $$\sup \left\{ \frac{g^2(\vartheta)}{\int \vartheta^2 \, \mathrm{d}\xi} : \int \vartheta^2 \, \mathrm{d}\xi \neq 0,\ \vartheta \in \Theta \right\} < \infty. \tag{22}$$

In such a case

$$\mathrm{var}_\xi g = \lim_{n \to \infty} \int l_n^2 \, \mathrm{d}\xi = \int (P_\xi l)^2 \, \mathrm{d}\xi$$

$$= \sup \left\{ \frac{g^2(\vartheta)}{\int \vartheta^2 \, \mathrm{d}\xi} : \int \vartheta^2 \, \mathrm{d}\xi \neq 0,\ \vartheta \in \Theta \right\}. \tag{23}$$

PROOF. According to Eq. (20), g is estimable under ξ exactly if

$$\int l\vartheta \, \mathrm{d}\xi = \mathrm{E}_\vartheta\left[\int l(x) y^\xi(\mathrm{d}x)\right] = g(\vartheta); \quad (\vartheta \in \Theta)$$

for some $l \in L^2(\xi)$. Since $\int l\vartheta \, \mathrm{d}\xi = \langle l, \vartheta \rangle_\xi$, we obtain

$$\int (P_\xi l)\vartheta \, \mathrm{d}\xi = \int l\vartheta \, \mathrm{d}\xi; \quad (\vartheta \in \Theta).$$

According to Eq. (21), we have

$$D[\int l(x)y^{\xi}\,(dx)] = \int l^2(x)\xi\,(dx),$$

hence the BLUE of g is

$$\hat{g} = \int (P_\xi l)(x)y^{\xi}\,(dx),$$

and

$$\text{var}_\xi g = \int (P_\xi l)^2\,d\xi.$$

Since $P_\xi l \in \bar{\Theta}^{\xi}$, there are $l_1, l_2, \ldots \in \Theta$ such that

$$\lim_{n\to\infty} \|l_n - P_\xi l\|_\xi^2 = 0.$$

Hence

$$\lim_{m,\,n\to\infty} \|l_n - l_m\|_\xi^2 = 0,$$

$$\lim_{n\to\infty} \int l_n \vartheta\,d\xi = \int (P_\xi l)\vartheta\,d\xi; \quad (\vartheta \in \Theta).$$

Consequently, a)⇒b).

We have

$$\int (P_\xi l)^2\,d\xi = \lim_{n\to\infty} \frac{[\int (P_\xi l) l_n\,d\xi]^2}{\int l_n^2\,d\xi}$$

$$\leqslant \sup\left\{\frac{g^2(\vartheta)}{\int \vartheta^2\,d\xi} : \vartheta \neq 0,\ \vartheta \in \Theta\right\}. \tag{24}$$

On the other hand, from the Schwarz inequality in $L^2(\xi)$ it follows that

$$g^2(\vartheta) = [\int (P_\xi l)\vartheta\,d\xi]^2 \leqslant \int (P_\xi l)^2\,d\xi \int \vartheta^2\,d\xi. \tag{25}$$

Comparing the expressions in Eqs. (24) and (25), we obtain that b)⇒c) and that there holds the inequality (22).

Finally, from c) it follows that g is continuous on Θ with respect to $\|\ \|_\xi$, hence, according to the Riesz theorem, there is an $l \in L^2(\xi)$ such that

$$g(\vartheta) = \langle l, \vartheta \rangle_\xi; \quad (\vartheta \in \Theta).$$

Thus c)⇒a). □

PROPOSITION VII.3 [85]. *A functional g is estimable at least under one design exactly if*

$$\sup\left\{\frac{|g(\vartheta)|}{\|\vartheta\|_\infty}: 0\neq\vartheta\in\Theta\right\}<\infty. \tag{26}$$

In that case there is a finite generalized measure ν on $\mathcal{B}_0$ such that

$$g(\vartheta)=\int\vartheta\,\mathrm{d}\nu;\quad (\vartheta\in\Theta), \tag{27}$$

$$\sup\left\{\frac{|g(\vartheta)|}{\|\vartheta\|_\infty}: 0\neq\vartheta\in\Theta\right\}$$

$$=\sup\left\{\frac{|\int l\,\mathrm{d}\nu|}{\|l\|_\infty}: 0\neq l\in C(X)\right\}. \tag{28}$$

Let $\nu=\nu^+-\nu^-$ be the Jordan decomposition of ν. Then

$$\xi_\nu\equiv(\nu^++\nu^-)/[\nu^+(X)+\nu^-(X)]$$

is optimal in the sense that

$$\mathrm{var}_{\xi_\nu}g=\min\{\mathrm{var}_\xi g: \xi\in\bar{\Xi}\}.$$

PROOF. Since $\int\vartheta^2\,\mathrm{d}\xi\leqslant\|\vartheta\|^2_\infty$, we have

$$\sup\left\{\frac{g^2(\vartheta)}{\|\vartheta\|^2_\infty}: 0\neq\vartheta\in\Theta\right\}$$

$$\leqslant\sup\left\{\frac{g^2(\vartheta)}{\int\vartheta^2\,\mathrm{d}\xi}: 0\neq\vartheta\in\Theta\right\}. \tag{29}$$

Hence, if g is estimable under ξ, then, according to Proposition VII.2, Eq. (26) holds.

Conversely, if Eq. (26) holds, then g is continuous on Θ with respect to the norm $\|\ \|_\infty$. According to the Hahn-Banach theorem (Chapter VII.1) there is a functional g^* defined on $C(X)$, such that

$$g(\vartheta)=g^*(\vartheta);\quad (\vartheta\in\Theta)$$

and that, according to Eq. (10),

$$\sup\left\{\frac{g^2(\vartheta)}{\|\vartheta\|^2_\infty}: 0\neq\vartheta\in\Theta\right\}=\sup\left\{\frac{[g^*(l)]^2}{\|l\|^2_\infty}: 0\neq l\in C(X)\right\}.$$

According to the Riesz representation theorem (Chapter VII.1) there is a finite signed measure ν^* on the Borel σ-algebra $\mathscr{B}$, such that $g^*(\cdot)=\int(\cdot)\, d\nu^*$. Let ν be the restriction of ν^* to $\mathscr{B}_0$. Then

$$g(\vartheta)=\int \left(\frac{d\nu}{d\xi_\nu}\right) \vartheta \, d\xi_\nu\,; \quad (\vartheta\in\Theta), \tag{30}$$

hence g is estimable under ξ_ν (Proposition VII.2). Moreover,

$$\sup\left\{\frac{\left[\int l \frac{d\nu}{d\xi_\nu}\, d\xi_\nu\right]^2}{\|l\|_\infty^2} : 0\neq l\in C(X)\right\}$$

$$\geqslant\left(\frac{d\nu}{d\xi_\nu}\right)^2=\int\left(\frac{d\nu}{d\xi_\nu}\right)^2 d\xi_\nu \geqslant \operatorname{var}_{\xi_\nu} g, \tag{31}$$

the last inequality following from Eq. (30) and from Proposition VII.2. Comparing Eq. (31) with Eq. (29) we obtain

$$\operatorname{var}_{\xi_\nu} g \leqslant \operatorname{var}_\xi g\,; \quad (\xi\in\bar{\Xi}). \ \square$$

Denote by G the linear space of linear functionals (defined on Θ) important to the experimenter. Suppose that $\dim G=r<\infty$. Let $g_1, \ldots, g_r$ be a basis in G. Then $g\in G$ implies

$$g=\sum_{i=1}^{r} \beta_i g_i \tag{32}$$

for some (unique) $\beta_1, \ldots, \beta_r$.

Proposition VII.4. *The functionals $g_1, \ldots, g_r$ are estimable under ξ exactly if every functional $g\in G$ is estimable under ξ. In that case the covariance matrix $\mathbf{D}_G$ of the BLUE-s $\hat{g}_1, \ldots, \hat{g}_r$ is nonsingular.*

Every functional g_i is estimable at least under one design exactly if every $g\in G$ is estimable at least under one design.

Proof. The first statement follows evidently from Eq. (32) and from Proposition VII.2, part a). From Eq. (23) we obtain

$$\{\mathbf{D}_G(\xi)\}_{ij}=\int (P_\xi l_i)(P_\xi l_j)\, d\xi,$$

where $l_1, \ldots, l_r\in L^2(\xi)$ are defined by

$$g_i(\vartheta) = \int l_i\vartheta \, d\xi; \quad (\vartheta \in \Theta).$$

Obviously,

$$\mathbf{u}'\mathbf{D}_G(\xi)\mathbf{u} = \int \left[\sum_{i=1}^{r} u_i P_\xi l_i\right]^2 d\xi > 0$$

for every $\mathbf{u} = (u_1, \ldots, u_r)$, $\mathbf{u} \neq \mathbf{0}$.

The last statement follows from Proposition VII.3. □

The inequality

$$\mathbf{D}_G(\xi) \leq \mathbf{D}_G(\eta)$$

is equivalent to

$$\mathrm{var}_\xi g \leq \mathrm{var}_\eta g; \quad (g \in G),$$

hence, as in the regression model, the uniform ordering of designs can be considered (see Chapter III.2) relative to the variances of $\hat{g}$; $(g \in G)$. Similarly, various optimality criteria can be defined:

D-optimality. It is defined by the criterion function

$$\xi \in \bar{\Xi} \mapsto \begin{cases} \log \det \mathbf{D}_G(\xi) & \text{if } g_1, \ldots, g_r \text{ are estimable,} \\ \infty, & \text{otherwise.} \end{cases}$$

L_p-criteria. The criteria functions are

$$\xi \in \bar{\Xi} \mapsto \begin{cases} \left[\frac{1}{r} \operatorname{tr} (\mathbf{H}'\mathbf{D}_G(\xi)\mathbf{H})^p\right]^{1/p} & \text{if } g_1, \ldots, g_r \text{ are estimable,} \\ \infty, & \text{otherwise.} \end{cases}$$

In the following two propositions an extremal situation is considered when Θ is dense in $C(X)$. The reader is referred to [86] for the proofs of the propositions.

PROPOSITION VII.5 [86]. *Let G be generated by r linearly independent functionals $g_1, \ldots, g_r$ defined on the set Θ that is dense in $C(X)$. Suppose that $g_1, \ldots, g_r$ are estimable under a design $\varkappa$. Let $l_1, \ldots, l_r \in L^2(\varkappa)$ be such that*

$$g_i(\vartheta) = \int l_i\vartheta \, d\varkappa; \quad (\vartheta \in \Theta, i = 1, \ldots, r).$$

Let $\mathcal{B}_\varkappa \subset \mathcal{B}_0$ be the minimum σ-algebra ensuring the measurability of the functions $l_1, \ldots, l_r$. Then given any design ξ, there is a design μ such that

a) $\frac{d\mu}{d\varkappa}$ *exists and is $\mathcal{B}_\varkappa$-measurable,*

b) $g_1, \ldots, g_r$ *are estimable under μ,*

c) *for every $g \in G$ we have $\mathrm{var}_\mu g \leqslant \mathrm{var}_\xi g$.*

Suppose now that the design $\varkappa$ from Proposition VII.5 can be chosen so that $\mathcal{B}_\varkappa$ is a finite algebra. Denote the $\varkappa$-atoms of $\mathcal{B}_\varkappa$ by $\{A_i\}_{i=1}^k$. (Note that A_i is a $\varkappa$-atom of $\mathcal{B}_\varkappa$ iff $\varkappa(A_i)>0$, $B \in \mathcal{B}_\varkappa$, $B \subsetneqq A_i$ imply $\varkappa(B)=0$.) In such a case any design ξ is defined by the numbers

$$\xi(A_1), \ldots, \xi(A_k).$$

PROPOSITION VII.6. *Under the assumptions in Proposition VII.5 there is a unique D-optimum design μ.*

Let $\{\mathbf{u}^{(n)}\}_{n=0}^\infty$ be a sequence of vectors that belong to the simplex

$$\left\{\mathbf{u}: \mathbf{u} \in R^k,\ u_i \geqslant 0, \sum_{i=1}^k u_i \leqslant 1\right\} \tag{33}$$

and are obtained by the algorithm:

a) $$u_i^{(0)}>0; \quad (i=1, \ldots, k), \ \sum_{i=1}^k u_i^{(0)}=1,$$

b) $$u_i^{(n+1)}=\left[\frac{1}{r}\sum_{j,l=1}^r v_j(A_i)\{\mathbf{D}_G^{-1}(\mathbf{u}^{(n)})\}_{jl} v_l(A_i)\right]^{1/2}, \tag{34}$$

where

$$\{\mathbf{D}_G(\mathbf{u}^{(n)})\}_{jl}=\sum_{i=1}^k \frac{v_j(A_i)v_l(A_i)}{u_i^{(n)}}; \quad (j, l=1, \ldots, r). \tag{35}$$

Then

$$\lim_{n\to\infty} u_i^{(n)}=\mu(A_i); \quad (i=1, \ldots, k)$$

and there holds the stopping rule

$$\sum_{i=1}^k \frac{[u_i^{(n)}-\mu(A_i)]^2}{\mu(A_i)} \leqslant 2\left[1-\sum_{i=1}^k u_i^{(n)}\right]. \tag{36}$$

VII.4. EXPERIMENTS FOR THE ESTIMATION OF NONLINEAR FUNCTIONALS

Consider the functional model $(X, \Theta, 1)$ of the order zero again. Suppose the observed random variables $\{y^{\xi}(B): B\in\mathcal{B}_0\}$ to be Gaussian, i.e.

$$P_{\vartheta}\{\omega: y^{\xi}(B, \omega)<c\}$$
$$=[2\pi\xi(B)]^{-1/2}\int_{-\infty}^{c} \exp\left\{-\frac{(t-\int \vartheta\chi_B\,d\xi)^2}{2\xi(B)}\right\} dt. \tag{37}$$

When using the tensor product of Hilbert spaces and the properties of the Wiener chaos (cf. [87]) one obtains unbiassed estimates of nonlinear functionals. We present such estimates when the functionals are homogeneous polynomials.

Let

$$p_n: \Theta\mapsto R$$

be a homogeneous polynomial of n-th degree. That means

$$p_n(\vartheta)=h_n(\vartheta, ..., \vartheta); \quad (\vartheta\in\Theta) \tag{38}$$

for some multilinear functional

$$h_n: \Theta^n\mapsto R.$$

Particularly, when considering the regression model

$$\Theta=\left\{\sum_{i=1}^{m} \alpha_i f_i: \alpha\in R^m\right\},$$

p_n is a polynomial of n-th degree in the variables $\alpha_1, ..., \alpha_m$ (see Proposition VII.12).

EXAMPLES

1. $p_n(\vartheta)=\vartheta^n(x)$,

where $x\in X$ is a fixed point.

2. $p_n(\vartheta)=\sum_{k=0}^{n} c_k\vartheta^k(x_1)\vartheta^{n-k}(x_2)$,

where $x_1, x_2 \in X$ are fixed.

3. $p_n(\vartheta) = \sum_{k=0}^{n} c_k \int \vartheta^k \, d\xi \int \vartheta^{n-k} \, d\eta,$

where ξ, η are measures on $\mathscr{B}_0$, etc.

Denote by $\vartheta_1 \otimes \ldots \otimes \vartheta_n$ the real-valued function defined on X^n by

$$\vartheta_1 \otimes \ldots \otimes \vartheta_n(x_1, \ldots, x_n) \equiv \prod_{i=1}^{n} \vartheta_i(x_i) \tag{39}$$

(the tensor product of functions).

Further denote

$$\vartheta^{\otimes n} \equiv \vartheta \otimes \ldots \otimes \vartheta$$

(the tensor power).

The n-th tensor power of Θ, $\Theta^{\otimes n}$, is the linear space of functions on X^n spanned by the set

$$\{\vartheta_1 \otimes \ldots \otimes \vartheta_n : \vartheta_1, \ldots, \vartheta_n \in \Theta\}.$$

The symmetric n-th tensor power, $\Theta^{\odot n}$, is the linear space spanned by

$$\{\vartheta^{\otimes n} : \vartheta \in \Theta\}.$$

Clearly, $\Theta^{\odot n}$ is the linear space of all symmetric functions belonging to $\Theta^{\otimes n}$. By the symmetry we understand the invariance to any permutation of the coordinates of the points $(x_1, \ldots, x_n) \in X^n$.

As known, there is a unique linear functional q_n defined on $\Theta^{\odot n}$ and such that

$$q_n(\vartheta^{\otimes n}) = p_n(\vartheta); \quad (\vartheta \in \Theta). \tag{40}$$

In what follows, designs for the estimation of p_n in the experiment $(X, \Theta, 1)$ will be compared with designs for the estimation of the linear functional q_n in a modified experiment $(X^n, \Theta^{\odot n}, 1)$.

Let $\xi \in \bar{\Xi}$ be a design in $(X, \Theta, 1)$. The polynomial p_n is estimated by a polynomial in the variable $y^{\xi}(B)$; $(B \in \mathscr{B}_0)$ or by a limit of such polynomials. Denote by $\mathscr{P}_{\xi}$ the subspace of $L^2(\Omega, \mathscr{S}, P_0)$ spanned by all polynomials in the variables $y^{\xi}(B)$; $(B \in \mathscr{B}_0)$.

If $y \in \mathscr{P}_\xi$, then, as proven in Proposition VII.9,

$$\|y\|_0^2 \leq \|y\|_\vartheta^2.$$

Hence, in case of $\vartheta \neq 0$, the subspace of $L^2(\Omega, \mathscr{S}, P_\vartheta)$ spanned by the polynomials in $y^\xi(B)$; $(B \in \mathscr{B}_0)$ is a subset of $\mathscr{P}_\xi$ but it is not equal to $\mathscr{P}_\xi$. In other words, not all random variables belonging to $\mathscr{P}_\xi$ have a finite variance under the hypothesis $\vartheta \neq 0$.

A random variable $y \in \mathscr{P}_\xi$ is an unbiassed estimate of p_n if

$$D_\vartheta(y) \equiv \int [y - E_\vartheta(y)]^2 \, dP_\vartheta < \infty$$

for every $\vartheta \in \Theta$, and if

$$E_\vartheta(y) = p_n(\vartheta); \quad (\vartheta \in \Theta). \tag{41}$$

For further use, notice that a random variable y is Gaussian with the mean $E(y)$ and the variance $D(y)$ exactly if

$$E[e^{uy}] = e^{uE(y)} e^{(1/2)u^2 D(y)}; \quad (u \in R) \tag{42}$$

(cf. [87]).

Denote by $\mathscr{S}_y$ the smallest sub-σ-algebra of $\mathscr{S}$ ensuring the measurability of the functions

$$\omega \in \Omega \mapsto y^\xi(B, \omega); \quad (B \in \mathscr{B}_0)$$

(usually $\mathscr{S} = \mathscr{S}_y$).

PROPOSITION VII.7. *Let $\vartheta \in \Theta$. The probability distributions P_0 and P_ϑ are equivalent on $\mathscr{S}_y$ (i.e. $P_0 \ll P_\vartheta$, and $P_\vartheta \ll P_0$), and their derivative dP_ϑ/dP_0 on $\mathscr{S}_y$ is equal to*

$$\frac{dP_\vartheta}{dP_0}(\omega) = \exp\left\{\int \vartheta(x) y^\xi(dx, \omega) - \frac{1}{2}\int \vartheta^2 \, d\xi\right\}. \tag{43}$$

PROOF [87]. Define a measure P on $\mathscr{S}_y$ by

$$P \equiv \exp\left\{\int \vartheta(x) y^\xi(dx) - \frac{1}{2}\int \vartheta^2 \, d\xi\right\} P_0.$$

From Eq. (42) it follows that

$$P(\Omega) = \exp\left\{-\frac{1}{2}\int \vartheta^2 \, d\xi\right\} E_{P_0}[\exp\{\int \vartheta(x) y^\xi(dx)\}] = 1,$$

since according to Eq. (20),

$$\mathrm{E}_{P_0}[\int \vartheta(x) y^{\xi}(\mathrm{d}x)] = \mathrm{E}_0[\Psi(\vartheta)] = 0,$$

and, according to Eq. (21),

$$\mathrm{D}_{P_0}[\int \vartheta(x) y^{\xi}(\mathrm{d}x)] = \mathrm{D}_0[\Psi(\vartheta)] = \int \vartheta^2 \,\mathrm{d}\xi.$$

Obviously, $(\mathrm{d}P/\mathrm{d}P_0)(\omega) > 0$; $(\omega \in \Omega)$, hence $P_0 \ll P$. Further

$$\begin{aligned}
\mathrm{E}_P(\mathrm{e}^{uy^{\xi}(B)}) &= \mathrm{E}_0[\exp\{\int (u\chi_B(x) + \vartheta(x)) y^{\xi}(\mathrm{d}x)\}] \exp\left\{-\frac{1}{2}\int \vartheta^2 \,\mathrm{d}\xi\right\} \\
&= \exp\left\{\frac{1}{2}\int (u\chi_B + \vartheta)^2 \,\mathrm{d}\xi\right\} \exp\left\{-\frac{1}{2}\int \vartheta^2 \,\mathrm{d}\xi\right\} \\
&= \exp\left\{u \int \chi_B \vartheta \,\mathrm{d}\xi + \frac{1}{2} u^2 \xi(B)\right\}.
\end{aligned}$$

When compared with Eq. (42) one can see that $y^{\xi}(B)$ is Gaussian with the mean $\int \chi_B \vartheta \,\mathrm{d}\xi$ and the variance $\xi(B)$. It follows that $P = P_\vartheta$ on $\mathcal{S}_y$. □

Denote by ξ^n the measure on $(X^n, \mathcal{B}_0^n)$ defined by

$$\xi^n(B_1 \times \ldots \times B_n) = \prod_{i=1}^{n} \xi(B_i); \quad (B_1, \ldots, B_n \in \mathcal{B}_0). \tag{44}$$

Denote by $[L^2(\xi)]^{\odot n}$ the subspace of $L^2(X^n, \mathcal{B}_0^n, \xi^n)$ spanned by

$$\{l^{\otimes n} : l \in L^2(\xi)\}.$$

The Hilbert space $[L^2(\xi)]^{\odot n}$ is said to be the n-th symmetric tensor power of $L^2(\xi)$. In particular, the 0-th symmetric tensor power, $[L^2(\xi)]^{\odot 0}$, is identified with R.

Denote by

$$\exp \odot L^2(\xi) \equiv \bigoplus_{i=0}^{\infty} [L^2(\xi)]^{\odot i} \tag{45}$$

the direct sum of the symmetric tensor powers. Finally, denote

$$\exp \odot l \equiv 1 + \sum_{i=1}^{\infty} \frac{l^{\otimes i}}{\sqrt{i!}}; \quad (l \in L^2(\xi)). \tag{46}$$

Connection between the Hilbert spaces $\exp \odot L^2(\xi)$ and $\mathcal{P}_\xi$ is expressed in the following proposition.

PROPOSITION VII.8 [87]. *The set*

$$\{\exp \odot l\colon l \in L^2(\xi)\}$$

is a linearly independent system which spans the Hilbert space $\exp \odot L^2(\xi)$.

Correspondingly, the set

$$\left\{\exp\left\{\int l(x)y^\xi\,(\mathrm{d}x) - \frac{1}{2}\int l^2\,\mathrm{d}\xi\right\} : l \in L^2(\xi)\right\}$$

is a linearly independent system of random variables which spans the Hilbert space $\mathcal{P}_\xi$, *under the hypothesis* $\vartheta = 0$.

There is a linear bijection φ *of* $\exp \odot L^2(\xi)$ *onto* $\mathcal{P}_\xi$, *defined by*:

$$\varphi\,(\exp \odot l) = \exp\left\{\int l(x)y^\xi\,(\mathrm{d}x) - \frac{1}{2}\int l^2\,\mathrm{d}\xi\right\}; \qquad (47)$$

$$(l \in L^2(\xi)).$$

Under the hypothesis $\vartheta = 0$, *the mapping* φ *preserves the inner product*

$$\int_\Omega \varphi(h_1)\varphi(h_2)\,\mathrm{d}P_0 = \langle h_1, h_2\rangle; \qquad (48)$$

$$(h_1, h_2 \in \exp \odot L^2(\xi)).$$

The proof of the proposition is given in [87]. It is based on the equality

$$\int_\Omega \varphi\,(\exp \odot l_1)\varphi\,(\exp \odot l_2)\,\mathrm{d}P_0 = \langle \exp l_1, \exp l_2\rangle; \qquad (49)$$

$$(l_1, l_2 \in L^2(\xi)),$$

which is a particular case of Eq. (48), and which will be proven now.

From Eq. (46) it follows that

$$\langle \exp \odot l_1, \exp \odot l_2\rangle = 1 + \sum_{i=1}^{\infty} \frac{[\int l_1 l_2\,\mathrm{d}\xi]^i}{i!} = \exp\{\textstyle\int l_1 l_2\,\mathrm{d}\xi\}. \qquad (50)$$

On the other hand, using Eq. (42) we obtain

$$\mathrm{E}_0\left[\exp\left\{\int l_1(x)y^{\xi}(\mathrm{d}x)-\frac{1}{2}\int l_1^2\,\mathrm{d}\xi\right\}\exp\left\{\int l_2(x)y^{\xi}(\mathrm{d}x)-\frac{1}{2}\int l_2^2\,\mathrm{d}\xi\right\}\right]$$

$$=\exp\left\{\frac{1}{2}\int(l_1+l_2)^2\,\mathrm{d}\xi-\frac{1}{2}\int l_1^2\,\mathrm{d}\xi-\frac{1}{2}\int l_2^2\,\mathrm{d}\xi\right\}=\exp\left\{\int l_1 l_2\,\mathrm{d}\xi\right\}. \quad (51)$$

Thus Eq. (49) is obtained after comparing Eq. (50) with Eq. (51). □

PROPOSITION VII.9. [71]. *If $y\in\mathcal{P}_\xi$ is an unbiassed estimate of p_n, then there is an $l_y\in[L^2(\xi)]^{\odot n}$ such that*

$$y=\varphi(l_y),$$

$$p_n(\vartheta)=[n!]^{-1/2}\int_{X^n} l_y(x_1,\ldots,x_n)\vartheta(x_1)\ldots\vartheta(x_n)\xi(\mathrm{d}x_1)\ldots\xi(\mathrm{d}x_n)$$

for every $\vartheta\in\Theta$. Finally, the variance of y under the hypothesis $\vartheta=0$ is

$$\mathrm{D}_0(y)=\int l_y^2(x_1,\ldots,x_n)\xi(\mathrm{d}x_1)\ldots\xi(\mathrm{d}x_n).$$

PROOF. According to Proposition VII.7 there is an

$$l\equiv\bigoplus_{i=0}^{\infty} l_i\in\exp\odot L^2(\xi)$$

such that $y=\varphi(l)$. From Propositions VII.7 and VII.8 and from Eq. (40) we obtain

$$\begin{aligned}q_n(\vartheta^{\otimes n})&=p_n(\vartheta)=\mathrm{E}_\vartheta(y)\\&=\int_\Omega y\exp\left\{\int\vartheta(x)y^{\xi}(\mathrm{d}x)-\frac{1}{2}\int\vartheta^2\,\mathrm{d}\xi\right\}\mathrm{d}P_0\\&=\langle l,\exp\odot\vartheta\rangle\\&=\sum_{i=0}^{\infty}\int_{X^i} l_i\frac{\vartheta^{\otimes i}}{\sqrt{i!}}\,\mathrm{d}\xi^i;\quad(\vartheta\in\Theta).\end{aligned}$$

Since q_n is a linear function of $\vartheta^{\otimes n}$, we have $l_i=0$ for $i\neq n$. Thus

$$y=\varphi(l_n),$$

$$q_n(\vartheta^{\otimes n}) = p_n(\vartheta) = \frac{1}{\sqrt{n!}} \int_{X^n} l_n \vartheta^{\otimes n} \, d\xi^n ; \quad (\vartheta \in \Theta).$$

From Proposition VII.8 it follows that

$$D_0(y) = \|\varphi(l_n)\|_0^2 = \int_{X^n} l_n^2 \, d\xi^n . \square$$

PROPOSITION VII.10 [70]. *Given $l \in [L^2(\xi)]^{\odot n}$, there hold the inequalities*

$$\int l^2 \, d\xi^n \leqslant D_\vartheta[\varphi(l)] \leqslant \int l^2 \, d\xi^n \sum_{s=0}^{n-1} \binom{n}{s} \frac{[\int \vartheta^2 \, d\xi]^s}{s!} ;$$
$$(\vartheta \in \Theta).$$

PROOF. We shall use the Hermite polynomials H_i: $R \mapsto R$ defined by

$$H_i(t) = \frac{(-1)^i}{\sqrt{i!}} e^{t^2/2} \left(\frac{d}{dt}\right)^i e^{-(t^2/2)} \tag{52}$$

(cf. [97]). The last expression can be written as

$$H_i(t) = \frac{1}{\sqrt{i!}} \left(t - \frac{d}{dt}\right)^i 1. \tag{53}$$

Let $\{h_n\}_{n=1}^\infty$ be an orthogonal basis in $L^2(\xi)$. Obviously, the functions $h_{i_1} \otimes \ldots \otimes h_{i_n}$ are orthogonal and normed in $L^2(\xi^n)$. Denote by $\mathcal{N}$ the set of all sequences of nonzero integers $\{n_i\}_{i=1}^\infty$ such that $\sum_{i=1}^\infty n_i = n$. Let $n_{r_1}, \ldots, n_{r_k}$ be the nonzero terms in $\{n_i\} \in \mathcal{N}$. Denote

$$h(\{n_i\}) \equiv \frac{1}{\sqrt{k!}} \sum_\pi \frac{h_{\pi r_1}^{\otimes n_1}}{n_1!} \otimes \ldots \otimes \frac{h_{\pi r_k}^{\otimes n_k}}{n_k!},$$

the sum being taken over all permutations of $r_1, \ldots, r_k$. By combinatorial reasoning we get that the system

$$\{h(\{n_i\}) : \{n_i\} \in \mathcal{N}\}$$

is an orthogonal basis of $[L^2(\xi)]^{\odot n}$. Moreover, from the properties of φ

and Ψ (see Propositions VII.1 and VII.7) and from Eq. (52) it follows that

$$\varphi[h(\{n_i\})]=\prod_{i=1}^{k} H_{n_{ri}}[\Psi(h_{r_i})]=\prod_{i=0}^{\infty} H_{n_i}[\Psi(h_i)]. \tag{54}$$

Taking the coefficients $c(\{n_i\})$ so that

$$\sum_{\mathcal{N}} c^2(\{n_i\})<\infty$$

we obtain

$$l=\sum_{\mathcal{N}} c(\{n_i\})h(\{n_i\}),$$

$$D_\vartheta[\varphi(l)]=D_0\left\{\sum_{\mathcal{N}} c(\{n_i\})\prod_{i=1}^{\infty} H_{n_i}[\Psi(h_i)+\int h_i\vartheta\, d\xi]\right\}.$$

Denote the series in the brackets by $Z(l)$. Using Eq. (53) we obtain

$$Z(l)=\sum_{\mathcal{N}} c(\{n_i\})\prod_{i=1}^{\infty}\frac{1}{\sqrt{n_i!}}\left[\Psi(h_i)+\int h_i\vartheta\, d\xi-\frac{\partial}{\partial\Psi(h_i)}\right]^{n_i} 1.$$

Given a finite sequence of positive integers $i_1, \ldots, i_n$, define a sequence $\{n_i\}\in\mathcal{N}$ such that j is repeated n_j times in $(i_1, \ldots, i_n)$. Denote

$$b(i_1, \ldots, i_n)\equiv c(\{n_i\})\left[\prod_{i=1}^{\infty} n_i!/n!\right]^{1/2}.$$

Using the symmetry of $b(i_1, \ldots, i_n)$ we can write

$$Z(l)=\frac{1}{\sqrt{n!}}\sum_{(1,n)} b(i_1, \ldots, i_n)\prod_{s=1}^{n}\left[\Psi(h_{i_s})\right.$$
$$\left.-\frac{\partial}{\partial\Psi(h_{i_s})}+\int h_{i_s}\vartheta\, d\xi\right] 1,$$

where we have used the abbreviated notation

$$\sum_{(s,r)}\equiv\sum_{i_s, i_s+1, \ldots, i_r=1}^{\infty}.$$

After further rearrangements we obtain

$$Z(l)=\frac{1}{\sqrt{n!}}\sum_{r=0}^{n}\binom{n}{r}\sum_{(1,n)} b(i_1,\ldots,i_n)\prod_{s=1}^{r}\int h_{i_s}\vartheta\,d\xi$$
$$\times\prod_{t=r+1}^{n}\left[\Psi(h_{i_t})-\frac{\partial}{\partial\Psi(h_{i_t})}\right]1$$
$$=\frac{1}{\sqrt{n!}}\sum_{r=0}^{n}\binom{n}{r}\sum_{(r+1,n)}\left\{\left[\sum_{(1,r)} b(i_1,\ldots,i_n)\prod_{s=1}^{r}\int h_{i_s}\vartheta\,d\xi\right]\right.$$
$$\left.\times\left[\prod_{i=1}^{\infty} m_i!\right]^{1/2}\prod_{i=1}^{\infty}H_{m_i}[\Psi(h_i)]\right\},$$

the number i being repeated m_i times in the sequence $i_{r+1},\ldots,i_n$. (Clearly, $\sum_i m_i=n-r$.) Exploiting Eq. (53) we obtain

$$Z(l)=\sum_{r=0}^{n}\binom{n}{r}^{1/2}\varphi\left\{\sum_{(r+1,n)}\left[\sum_{(1,r)} b(i_1,\ldots,i_n)\right.\right.$$
$$\left.\left.\times\prod_{s=1}^{r}\int h_{i_s}\vartheta\,d\xi/\sqrt{r!}\right]h_{r+1}\otimes\ldots\otimes h_{i_n}\right\}.$$

Hence

$$D_0[Z(l)]=\sum_{r=0}^{n-1}\binom{n}{r}\sum_{(r+1,n)}\left[\sum_{(1,r)} b(i_1,\ldots,i_n)\right.$$
$$\left.\times\prod_{s=1}^{r}\int h_{i_s}\vartheta\,d\xi/\sqrt{r!}\right]^2. \tag{55}$$

In the last sum the term corresponding to $r=0$ is equal to $\|l\|^2$. Hence

$$D_\vartheta[\varphi(l)]\geqslant\|l\|^2=\int_{X^n} l^2\,d\xi^n. \tag{56}$$

From Eq. (55) we obtain

$$D_0[Z(l)]=\sum_{r=0}^{n-1}\frac{1}{r!}\binom{n}{r}\sum_{(r+1,n)}\left[\left\langle\sum_{(1,r)} b(i_1,\ldots,i_r)h_{i_1}\otimes\ldots\otimes h_{i_r},\vartheta^{\otimes r}\right\rangle\right]^2$$
$$\leqslant\sum_{r=0}^{n-1}\binom{n}{r}\sum_{(1,n)} b^2(i_1,\ldots,i_n)\|\vartheta\|_\xi^{2r}/r!$$
$$=\|l\|^2\sum_{r=0}^{n-1}\binom{n}{r}[\textstyle\int\vartheta^2\,d\xi]^r/r!\ \square$$

COROLLARY 1. *If*

$$p_n(\vartheta) = \int_{X^n} l_n \vartheta^{\otimes n} \, d\xi^n ; \quad (\vartheta \in \Theta)$$

for some $l_n \in [L^2(\xi)]^{\odot n}$, *then* p_n *is estimable under* ξ. (Compare with Proposition VII.8!)

COROLLARY 2. *If* $\hat{p}_n$ *is the minimum variance unbiassed estimate of* p_n, *then*

$$\begin{aligned} D_0(\hat{p}_n) &\leq D_\vartheta(\hat{p}_n) \\ &\leq D_0(\hat{p}_n) \sum_{s=0}^{n-1} \binom{n}{s} \frac{[\int \vartheta^2 \, d\xi]^s}{s!} . \end{aligned} \tag{57}$$

PROPOSITION VII.11 [70]. *Let* q_n *be a linear functional defined on* $\Theta^{\odot n}$ *by Eq. (40). The variance of the BLUE for* q_n *in the experiment* $(X^n, \Theta^{\odot n}, 1)$ *under the design* ξ^n *is*

$$D(\hat{q}_n) = n! \, D_0(\hat{p}_n).$$

PROOF. From Proposition VII.2 we obtain

$$\begin{aligned} D(\hat{q}_n) = \min \Bigg\{ \int_{X^n} l^2 \, d\xi^n : \int_{X^n} l^2 \, d\xi^n < \infty, \\ \int_{X^n} lt \, d\xi^n = q_n(t); \quad (t \in \Theta^{\odot n}) \Bigg\} . \end{aligned} \tag{58}$$

According to Proposition VII.8, we have

$$\begin{aligned} n! \, D_0(\hat{p}_n) = n! \min \Bigg\{ \int_{X^n} l^2 \, d\xi^n : \int_{X^n} l^2 \, d\xi^n < \infty, \\ l \text{ is symmetric}, \quad [n!]^{1/2} \int_{X^n} l \vartheta^{\otimes n} \, d\xi^n = p_n(\vartheta); \quad (\vartheta \in \Theta) \Bigg\} . \end{aligned} \tag{59}$$

From the definition of the linear space $\Theta^{\odot n}$ it follows that

$$\int lt \, d\xi^n = q_n(t); \quad (t \in \Theta^{\odot n})$$

exactly if

$$\int_{X^n} l\vartheta^{\otimes n}\, d\xi^n = q_n(\vartheta^{\otimes n}); \quad (\vartheta \in \Theta).$$

Since the functions belonging to $\Theta^{\odot n}$ are symmetric, we can suppose in Eq. (58) that l is a symmetric function as well. Finally, $p_n(\vartheta) = q_n(\vartheta^{\otimes n})$; $(\vartheta \in \Theta)$. Hence the right-hand sides of Eqs. (58) and (59) are equal. □

The upper and the lower bounds of the variance $D_\vartheta[\hat{p}_n]$, given by Eq. (57), contain the term $D_0[\hat{p}_n]$ explicitly. This is the reason why we use $D_0[\hat{p}_n]$ as the value of an optimality criterion function, when a homogeneous polynomial p_n is to be estimated.

In case of $\dim \Theta = m < \infty$ (i.e. when the model is a regression model), we have the following.

PROPOSITION VII.12. *If*

$$\Theta = \left\{ \sum_{i=1}^{m} \alpha_i f_i : (\alpha_1, \ldots, \alpha_m)' \in R^m \right\},$$

then there are unique symmetric coefficients $\gamma(i_1, \ldots, i_n)$ *such that*

$$p_n(\vartheta) = \sum_{i_1, \ldots, i_n = 1}^{m} \gamma(i_1, \ldots, i_n) \alpha_{i_1} \ldots \alpha_{i_n}.$$

The polynomial p_n *is estimable under the design* ξ *exactly if there are coefficients* $\delta(j_1, \ldots, j_n)$; $(j_1, \ldots, j_n = 1, \ldots, m)$ *such that*

$$\gamma(i_1, \ldots, i_n) = \sum_{j_1, \ldots, j_n = 1}^{m} \{\mathbf{M}(\xi)\}_{i_1 j_1} \ldots \{\mathbf{M}(\xi)\}_{i_n j_n} \delta(j_1, \ldots, j_n).$$

There holds

$$D_0(\hat{p}_n) = \frac{1}{n!} \sum_{i_1, \ldots, i_n = 1}^{m} \sum_{j_1, \ldots, j_n = 1}^{m} \gamma(i_1, \ldots, i_n) \times \{\mathbf{M}^-(\xi)\}_{i_1 j_1} \ldots \{\mathbf{M}^-(\xi)\}_{i_n j_n} \gamma(j_1, \ldots, j_n).$$

PROOF. According to Proposition VII.11, $D_0(\hat{p}_n)$ is proportional to $D(\hat{q}_n)$. Hence we can use Proposition II.8 but in the case when q_n is estimated in the modified experiment $(X^n, \Theta^{\odot n}, 1)$. □

References

[1] Fisher, R. (1935), *The Design of Experiments*, Oliver Boyd, London.

[2] Kiefer, J. and Wolfowitz, J. (1959), Optimum Designs in Regression Problems, *Ann. Math. Statist.* **30**, 271—294.

[3] Klepikov, N. P. and Sokolov, S. N. (1964), *Analysis and Design of Experiments by the Maximum Likelihood Method*, Nauka, Moscow (in Russian).
Клепиков, Н. П. и Соколов, С. Н. (1964), *Анализ и планирование экспериментов методом максимума правдоподобия*, Наука, Москва.

[4] Fedorov, V. V. and Pázman, A. (1968), Design of Physical Experiments, *Fortschr. Phys.* **16**, 325—358.

[5] Nalimov, V. V. (1971), *Theory of Experiments*, Nauka, Moscow (in Russian).
Налимов, В. В. (1971), *Теория эксперимента*, Наука, Москва.

[6] Fedorov, V. V. (1972), *Theory of Optimal Experiments*, Academic Press, New York.

[7] Fedorov, V. V. and Malyutov, M. B. (1972), Optimal Designs in Regression Problems, *Math. Operationsforsch. Statist.* **3**, 281—308.

[8] Fellman, J. (1974), On the Allocation of Linear Observations, *Comment. Phys.-Math.* (Helsinki) **44**, 27—78.

[9] Nalimov, V. V. and Golikova, T. I. (1976), *Logical Foundations of Experimental Design*, Metalurgiya, Moscow (in Russian).
Налимов, В. В. и Голикова, Т. И. (1976), *Логические основания планирования экспериментов*, Металургия, Москва.

[10] Brodskii, V. V. (1976), *Introduction to the Design of Factorial Experiments*, Nauka, Moscow (in Russian).
Бродский, В. В. (1976), *Введение в факторное планирование эксперимента*, Наука, Москва.

[11] Bandemer, H. W. et al. (1977), *Theorie und Anwendung der optimalen Versuchsplanung* I. Akademie-Verlag, Berlin.

[12] Silvey, S. D. (1980), *Optimal Design*, Chapman Hall, London.

[13] Pázman, A. (1980), Some Features of the Optimal Design Theory — a Survey. *Math. Operationsforsch. Statist., Ser. Statist.* **11**, 415—446.

[14] Zarrop, M. B. (1979), *Optimal Experimental Design for Dynamic System Identifica-*

tion, Lecture Notes, Springer-Verlag, Berlin—Heidelberg—New York.

[15] Elfving, G. (1952), Optimum Allocation in Linear Regression, *Ann. Math. Statist.* **23**, 255—262.

[16] Kiefer, J. (1974), General Equivalence Theory for Optimum Designs (Approximate Theory), *Ann. Statist.* **2**, 849—879.

[17] Fedorov, V. V. and Pázman, A. (1967), Design of Experiments Based on the Measure of Information, Preprint JINR E5-3247, Dubna, USSR.

[18] Pázman, A. and Fedorov, V. V. (1967), Design of Specifying and Discriminating NN-scattering Experiments, *Nucl. Phys.* **6**, 853—859 (in Russian).
Пазман, А. и Федоров, В. В. (1967), Планирование уточняющих и дискриминирующих экспериментов по NN-рассеянию, *Ядер. физ.* **6**, 853—859.

[19] Nalimov, V. V. et al. (1977), *Regression Experiments (Design and Analysis)*, Moscow State University, Moscow (in Russian).
Налимов, В. В. и др. (1977), *Регрессионные эксперименты (планирование и анализ)*, Изд. МГУ, Москва.

[20] Golikova, T. I., Panchenko, L. A. and Fridman, M. Z. (1974), *Catalogue of Second-order Designs*, Moscow State University, Moscow (in Russian).
Голикова, Т. И. и Панченко, Л. А. и Фридман, М. З. (1974), *Каталог планов второго порядка*, Изд. МГУ, Москва.

[21] Gribik, P. R., Kortanek, K. O. and Sweigart, J. R. (1976), Designing a Regional Air Pollution Monitoring Network: An Appraisal of a Regression Experimental Design Approach, in EPA Conference on Modeling and Simulation, Cincinnati, Ohio.

[22] Kubáček, L. and Pázman, A. (1974), Optimum Nets, *Stud. Geophys. Geod.* **18**, 91—102.

[23] Mojžeš, M. (1975), Optimum Determination of Euler Angles, *Stud. Geophys. Geod.* **19**, 322—329.

[24] Pukelsheim, F. (1980), On Linear Regression Designs which Maximize Information, *J. Statist. Plann. Infer.* **4**, 339—364.

[25] Schoenberg, I. J. (1959), On the Maxima of Certain Hankel Determinants and the Zeros of the Classical Orthogonal Polynomials, *Nederl. Akad. Netensch., Proc. Ser. A62 — Indag. Math.* **21**, 282—290.

[26] Karlin, S. and Studden, W. J. (1966), Optimum Experimental Designs, *Ann. Math. Statist.* **37**, 783—815.

[27] Kiefer, J. (1961), Optimum Designs in Regression Problems, *Ann. Math. Statist.* **32**, 298—325.

[28] Box, G. E. P. (1952), Multifactor Designs of First Order, *Biometrika* **39**, 46—60.

[29] Raghavarao, D. (1971), Constructions and Combinatorial Problems in Design of Experiments, J. Wiley, New York.

[30] Hoel, P. G. (1965), Minimax Designs in Two-dimensional Regression, *Ann. Math. Statist.* **36**, 1097—1106.

[31] Hoel, P. (1965), Optimum Designs for Polynomial Extrapolations, *Ann. Math. Statist.* **36**, 1483—1487.

[32] Studden, W. J. (1982), Optimal Designs for Weighted Polynomial Regression Using Canonical Moments, in *Statistical Decision Theory and Related Topics III*, Vol. 2, Academic Press, New York, pp. 335—349.

[33] Studden, W. J. (1980), D_s-optimal Designs for Polynomial Regression Using Continued Fractions, *Ann. Statist.* **8**, 1132—1141.

[34] Gorskii, V. G. and Brodskii, V. V. (1969), First-order Simplex Designs and Related Second-order Designs, in V. V. Nalimov (Ed.), *New Ideas in Experimental Design*, Nauka, Moscow (in Russian).

Горский, В. Г. и Бродский, В. В. (1969), О сиплекс планах первого порядка и связанных с ними планов второго порядка, в В. В. Налимов (Ред.), *Новые идеи и планировании эксперимента*, Наука, Москва.

[35] Kiefer, J. (1975), Construction and Optimality of Generalized Youden Designs, in J. Srivastava (Ed.), *A Survey of Statistical Designs and Linear Model*, North-Holland, New York, pp. 333—353.

[36] Tocher, K. D. (1952), A Note on the Design Problem, *Biometrika* **39**, 189—195.

[37] Giovagnoli, A. and Wynn, H. P. (1981), Optimum Continuous Block Designs, *Proc. Roy. Soc. Lond.* **A377**, 405—416.

[38] Giovagnoli, A. and Wynn, H. P. (1983), Schur-optimal Continuous Block Designs, Manuscript.

[39] Pázman, A. (1966), The Analysis of Errors from the Linear Approximation in Data Performance, *Avtometriya* **2**, 76—84 (in Russian).

Пазман, А. (1966), Анализ погрешеностей линейного приближения при статистической обработке результатов измерения, *Автометрия* **2**, 76—84.

[40] O'Hagan, A. (1978), Curve Fitting and Optimal Designs for Prediction, *J. Roy. Statist. Soc.* **B40**, 1—42.

[41] Läuter, H. (1974), On the Admissibility and Nonadmissibility of the Usual Estimator for the Mean of Multivariate Normal Population and Conclusions to Optimal Design, *Math. Operationsforsch. Statist., Ser. Statist.* **5**, 591—597.

[42] Atwood, C. L. (1969), Optimal and Efficient Designs of Experiments, *Ann. Math. Statist.* **40**, 1570—1602.

[43] Atwood, C. L. (1976), Convergent Design Sequences for Sufficiently Regular Optimality Criteria, *Ann. Statist.* **4**, 1124—1138.

[44] Atwood, C. L. (1976), Convergent Design Sequences for Sufficiently Regular Optimality Criteria — Singular Case, Preprint Univ. of California, Davis.

[45] Pázman, A. (1972), Sequential Designs for Estimatings out of k Parameters, *Acta Metronomica* 4, Bratislava.

[46] Pázman, A. (1974), A Convergence Theorem in the Theory of D-optimum Experimental Designs, *Ann. Statist.* **2**, 216—218.

[47] Titterington, D. M. (1976), Algorithms for Computing D-optimal Designs on a Finite Design Space, in Proceedings of the 1976 Conference on Information Science and Systems, Department of Electronic Engineering, Johns Hopkins University, Baltimore, pp. 213—216.

[48] Atwood, C. L. (1973), Sequences Converging to D-optimal Experiments, *Ann. Statist.* **1**, 342—352.

[49] Sibson, R. (1972), D_A-optimality and Duality, in Proceedings of European Meeting of Statisticians, Budapest.

[50] Sibson, R. and Kenny, A. (1975), Coefficients in D-optimal Experimental Design, *J. Roy. Statist. Soc.* **B37**, 288—292.

[51] Silvey, S. D. and Titterington, D. M. (1973), A Geometric Approach to Optimal Design Theory, *Biometrika* **60**, 21—32.

[52] Silvey, S. D. (1974), Some Aspects of the Theory of Optimal Linear Regression Design with a General Concave Criterion Function, Techn. Rep. No. 75, Ser. 2, Department of Statistics, Princeton University, Princeton.

[53] Pukelsheim, F. (1981), On c-optimal Design Measures, *Math. Operationsforsch. Statist., Ser. Statist.* **12**, 13—20.

[54] Titterington, D. M. and Pukelsheim, F. (1983), General Differential and Lagrangian Theory for Optimal Experimental Design, *Ann. Statist.* **11**, 1060—1068.

[55] Titterington, D. M. (1975), Optimal Design: Some Geometrical Aspects of D-optimality, *Biometrika* **62**, 313—320.

[56] Whittle, P. (1973), Some General Points in the Theory of Optimal Experimental Design, *J. Roy. Statist. Soc.* **B35**, 123—130.

[57] Wynn, H. P. (1972), Results in the Theory and Construction of D-optimal Experimental Designs, *J. Roy. Statist. Soc.* **B34**, 133—147.

[58] Silvey, S. D., Titterington, D. M. and Torsney, B. (1978), An Algorithm for Optimal Designs on a Finite Design Space, *Commun. Statist.* **A**, 1379—1389.

[59] John, R. C. St. and Draper, N. R. (1975), D-optimality of Regression Design: A Review, *Technometrics* **17**, 15—23.

[60] Gaffke, N. (1983), Directional Derivatives of Optimality Criteria at Singular Matrices in Convex Design Theory, Manuscript.

[61] Gribik, P. R. and Kortanek, K. O. (1977), Equivalence theorems and Cutting Plane Algorithms for a Class of Experimental Design Problems, *SIAM J. Appl. Math.* **32**, 232—259.

[62] Wu, C. F. (1977), Characterizing the Consistent Directions for Least Squares Estimates, Techn. Rep. No. 506, Department of Statistics, University of Wisconsin, Madison.

[63] Wu, C. F. (1977), Some Algorithmic Aspects of the Theory of Optimal Designs, Techn. Rep. No. 499, Department of Statistics, University of Wisconsin, Madison.

[64] Wu, C. F. (1978), Some Iterative Procedures for Generating Nonsingular Optimal Designs, *Commun. Statist.* **A**, 1399—1412.

[65] Wu, C. F. and Wynn, H. P. (1978), The Convergence of General Steplength Algorithms for Regular Optimum Design Criteria, *Ann. Statist.* **6**, 1273—1285.

[66] Pázman, A. (1980), Singular Experimental Designs (Standard and Hilbert Space Approaches), *Math. Operationsforsch. Statist.* **11**, 137—149.

[67] Pázman, A. (1978), Computation of the Optimum Designs under Singular Information Matrices, *Ann. Statist.* **6**, 465—467.

[68] Mikulecká, J. (1983), Hybrid Experimental Design, *Kybernetika* (Prague) **19**, 1—14.

[69] Melas, V. B. (1978), Optimal Designs for Exponential Regression, *Math. Operationsforsch. Statist., Ser. Statist.* **4**, 45—59.

[70] Pázman, A. (1977), Plans d'expérience pour les estimations de fonctionelles non linéaires, *Ann. Inst. H. Poincaré* **13**, 259—267.

[71] Pázman, A. (1981), Optimal Designs for the Estimation of Polynomial Functionals, *Kybernetika* (Prague) **17**, 16—31.

[72] Volaufová, J. (1982), Estimation of Polynomials in the Regression Model, *Apl. Mat.* **27**, 223—231.

[73] Pázman, A. and Volaufová, J. (1982), Polynomials of Parameters in the Regression Model — Estimation and Design, in *Probability and Statistical Inference*, Reidel, Dordrecht, pp. 275—286.

[74] Pázman, A. (1984), Probability Density of Multivariate Nonlinear Least Squares Estimates, *Kybernetika* (Prague) **20**, 209—230.

[75] Sacks, J. and Ylvisaker, D. (1966), Designs for Regression with Autocorrelated Errors, *Ann. Math. Statist.* **37**, 66—89.

[76] Wahba, G. (1974), Regression Designs for Some Equivalence Classes of Kernels, *Ann. Statist.* **2**, 925—934.

[77] Hájek, J. and Kimeldorf, G. (1974), Regression Design in Autoregressive Stochastic Processes, *Ann. Statist.* **2**, 520—527.

[78] Kiefer, J. and Wynn, H. P. (1981), Optimum Balanced Block and Latin Square Designs for Correlated Observations, *Ann. Statist.* **4**, 737—757.

[79] Bickel, P. J. and Herzberg, A. M. (1979), Robustness of Design against Autocorrelation in Time, I, Asymptotic Theory, Optimality for Location and Linear Regression, *Ann. Statist.* **7**, 77—95.

[80] Hájek, J. (1962), On Linear Statistical Problems in Stochastic Processes, *Czechoslov. Math. J.* **12**, 486—491.

[81] Parzen, E. (1963), A New Approach to the Synthesis of Optimal Smoothing and Prediction Systems, *Math. Optim. Techn.* **10**, 75—108.

[82] Lauritzen, S. L. (1973), *The Probabilistic Background of Some Statistical Methods in Physical Geodesy*, Geodetic Institute, Kopenhagen.

[83] Kubáček, L. and Kubáčková, L. (1978), Filtration and Prediction of Inhomogeneous Anomalous Geophysical Potential Fields, *Stud. Geophys. Geod.* **4**, 330—336.

[84] Pázman, A. (1974), The Ordering of Experimental Designs: A Hilbert Space Approach, *Kybernetika* (Prague) **10**, 373—388.

[85] Pázman, A. (1978), Hilbert-space Methods in Experimental Design, *Kybernetika* (Prague) **14**, 73—84.

[86] Pázman, A. (1975—1976), Optimum Experimental Designs with a Lack of a priori Information, *Kybernetika* (Prague), Part 1: **11**, 355—367; Part 2: **12**, 7—14.

[87] Neveu, J. (1968), *Precessus aléatoires gaussiens*, Presses de l'Univ. de Montréal, Montréal.

[88] Rao, C. R. (1973), *Linear Statistical Inference and Its Applications*, J. Wiley, New York.

[89] Huber, P. J. (1973), Robust Regression: Asymptotic Conjectures and Monte Carlo, *Ann. Statist.* **1**, 799—821.

[90] Jurečková, J. (1980). Robust Estimation in a Linear Regression Model, in *Mathematical Statistics*, Banach Centre Publication, Vol. 6, pp. 169—174, Polish Sci. Publ., Warsawa.

[91] Jurečková, J. (1971), Nonparametric Estimates of Regression Coefficients, *Ann. Math. Statist.* **42**, 1328—1338.

[92] Bickel, P. V. (1973), On Some Analogues to Linear Combinations of Order Statistics in the Linear Model, *Ann. Statist.* **1**, 597—616.

[93] Thisted, R. A. (1976), Ridge Regression, Minimax Estimation, and Empirical Bayes Methods, Techn. Rep. No. 28, Stanford University, Stanford.

[94] Ferguson, T. S. (1967), *Mathematical Statistics, a Decision Theoretic Approach*, Academic Press, New York.

[95] Kubáček, L. and Pázman, A. (1979), *Štatistické metódy v meraní* (Statistical Methods in Measurements), VEDA, Bratislava.

[96] Cramér, H. (1946), *Mathematical Methods of Statistics*, Princeton University Press, Princeton.

[97] Szegö, G. (1975), *Orthogonal Polynomials*, 4th edition, American Mathematical Society, Providence.

[98] Kelley, J. L. (1955), *General Topology*, D. van Nostrand, New York.

[99] Kolmogorov, A. N. and Fomin, S. V. (1972), *Elements of the Theory of Functions and of the Functional Analysis*, 3rd edition, Nauka, Moscow (in Russian). Колмогоров, А. Н. и Фомин, С. В. (1972), *Элементы теории функций и функционального анализа*, 3. изд., Наука, Москва.

[100] Halmos, P. R. (1957), *Introduction to Hilbert Space*, 2nd edition, Chelsea Publ. Co., New York.

[101] Halmos, P. R. (1950), *Measure Theory*, D. van Nostrand, New York.

[102] Aronszajn, N. (1950), Theory of Reproducing Kernels, *Trans. Amer. Math. Soc.* **68**, 337—404.

[103] Rockafellar, R. T. (1970), *Convex Analysis*, Princeton University Press, Princeton.

[104] Holmes, R. B. (1975), *Geometric Functional Analysis and Its Applications*, Springer, New York.

[105] Kubáček, L. (1983), *Základy teórie odhadu* (Foundations of Estimation Theory), VEDA, Bratislava.

[106] Lankaster, P. (1969), *Theory of Matrices*, Academic Press, New York.

[107] Rao, C. R. and Mitra, S. K. (1971), *General Inverse of Matrices and Its Application*, J. Wiley, New York.

[108] Jarník, V. (1956), *Diferenciální počet II* (Differential Calculus II), 2nd edition, Publishing House of the Czechoslovak Academy of Sciences, Prague.

[109] Jarník, V. (1955), *Integrální počet II* (Integral Calculus II), Publishing House of the Czechoslovak Academy of Sciences, Prague.

[110] Belenkii, V. Z. et al. (1974), *Iterative Methods in Game Theory and in Programming*, Nauka, Moscow (in Russian).
Беленкий, В. З. и др. (1974), *Итеративные методы в теории игр и програмировании*, Наука, Москва.

[111] Marshall, A. W. and Olkin, I. (1979), *Inequalities: Theory of Majorization and Its Applications*, Academic Press, New York.

[112] Petrovskii, I. G. (1965), *Lectures on the Theory of Integral Equations*, Nauka, Moscow (in Russian).
Петровский, И. Г. (1965), *Лекции по теории интегральных уравнений*, Наука, Москва.

Subject Index